The School Mathematics Project

GCSE Mathematics Revision

Foundation Tier

PUBLISHED BY THE PRESS SYNDICATE OF THE UNIVERSITY OF CAMBRIDGE
The Pitt Building, Trumpington Street, Cambridge CB2 1RP, United Kingdom

CAMBRIDGE UNIVERSITY PRESS
The Edinburgh Building, Cambridge CB2 2RU, United Kingdom
40 West 20th Street, New York, NY 10011–4211, USA
10 Stamford Road, Oakleigh, Melbourne 3166, Australia

First published 1998

Printed in the United Kingdom at the University Press, Cambridge

A catalogue record for this book is available from the British Library

ISBN 0 521 57909 0

This book has been written and compiled by

Spencer Instone
Elizabeth Jackson
Susan Shilton

The authors' warm thanks go to Howard Baxter and William Wynne Willson who gave advice from an examiner's standpoint.

Certain questions in this book are reproduced by kind permission of the following:

The Midland Examining Group
The Northern Ireland Council for the Curriculum Examinations and Assessment
The Southern Examining Group
London Examinations: a division of Edexcel Foundation
(formerly University of London Examinations and Assessment Council)
The Welsh Joint Education Committee

These questions are acknowledged individually in the text. None of the above groups bears any responsibility for the accuracy or method of working in example answers to these questions.

The publishers would also like to thank David Marks Julia Barfield Limited for permission to reproduce the image of the British Airways Millennium Wheel on page 63 (image by Nick Wood/Hayes Davidson)

Contents

Shape, space and measures

Handling data

Mixed questions

Answers and hints

Formula sheet

Area of triangle $= \frac{1}{2} \times \text{base} \times \text{height}$

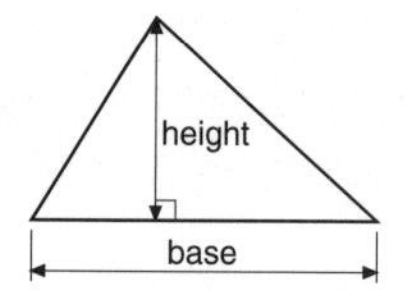

Circumference of circle $= \pi \times \text{diameter}$
$= 2 \times \pi \times \text{radius}$

Area of circle $= \pi \times (\text{radius})^2$

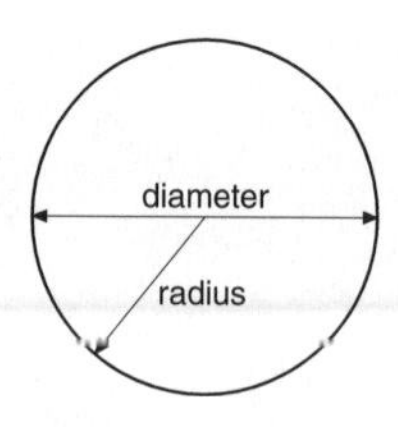

Volume of cuboid $= \text{length} \times \text{width} \times \text{height}$

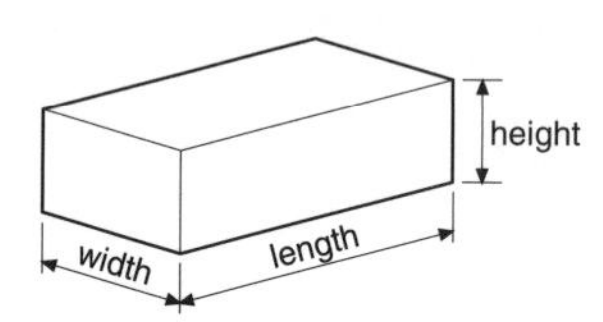

How to use this book

This book covers the content of all the GCSE mathematics syllabuses at foundation level. There is a brief explanation of each topic followed by questions (including many from past GCSE examinations) for you to work through. You will find answers and hints to all the questions at the back of the book.

How you use this book depends on how much you need to revise. It is divided into sections – Number, Algebra, Shape, space and measures and Handling data. You could start at the beginning of a section and work through it steadily or you could pick out the things that you are unsure about (with the help of the contents pages) and concentrate on them. If you need further help, references to SMP 11–16 books, at the end of each set of Answers and hints, tell you where to find it.

Some questions need worksheets. Ask your teacher for these.

NUMBER
Place value

The digit 2 in the number **2152672** has a different value in each of the three places it occurs.

2 152 672

Two millions

Two thousands

Two units

This number is two million, one hundred and fifty-two thousand, six hundred and seventy-two.

The same idea is true for decimal numbers.

Rounding ► page 4

1837·383

Three tens

Three tenths

Three thousandths

The decimal point separates the whole number part from the decimal part of the number.

1 What value is represented by the 4 in each of these?

(a) 5460 (b) 10·64 (c) 64550 (d) 8·41 (e) 0·014

2 Write in figures the numbers:

(a) four thousand and eight (b) twenty-six thousand one hundred and twenty

(c) three hundred and ten thousand (d) two hundred and one thousand

3 Write these numbers in order, starting with the smallest.

1100 988 1009 999 1080

4 Write these lengths in order of size, smallest first.

0·415 m 0·234 m 0·4 m 0·23 m 0·409 m

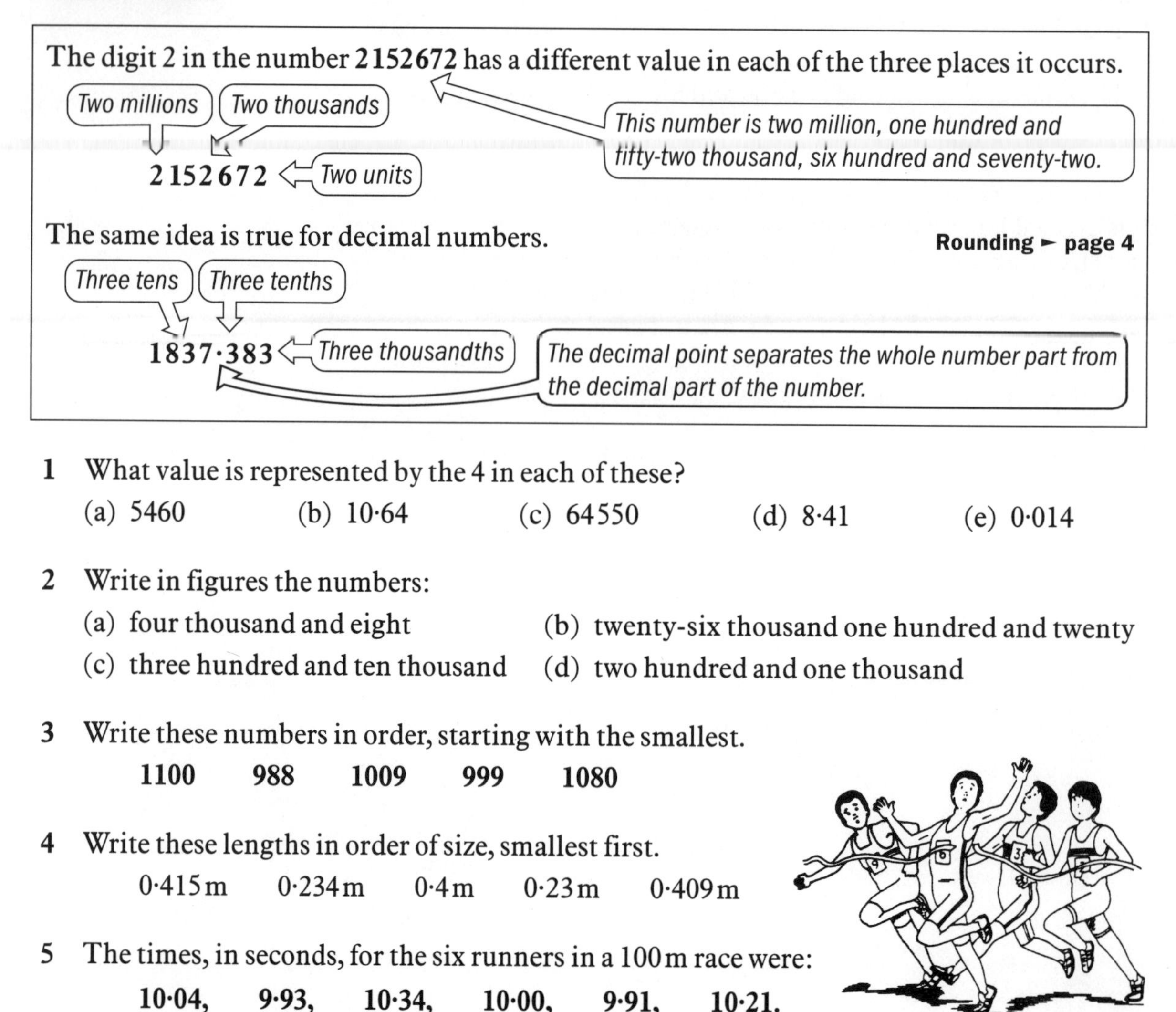

5 The times, in seconds, for the six runners in a 100 m race were:

10·04, 9·93, 10·34, 10·00, 9·91, 10·21.

What was the time of the winner?

MEG (SMP)

6 You can use these four number cards to make different numbers like 2816 and 2861.

Each card can be used only once and all the cards must be used.

(a) What is the **smallest** number you can make?

(b) What is the **largest** number you can make?

7 **5219 5912 5921 5291 5192 5129**

From the numbers above, write down the number which is:

(a) largest, (b) smallest, (c) nearest to 5200.

MEG/ULEAC (SMP)

Multiplying and dividing by tens, hundreds, thousands, ...

It is easy to multiply and divide whole numbers and decimals by 10s, 100s, 1000s, ...

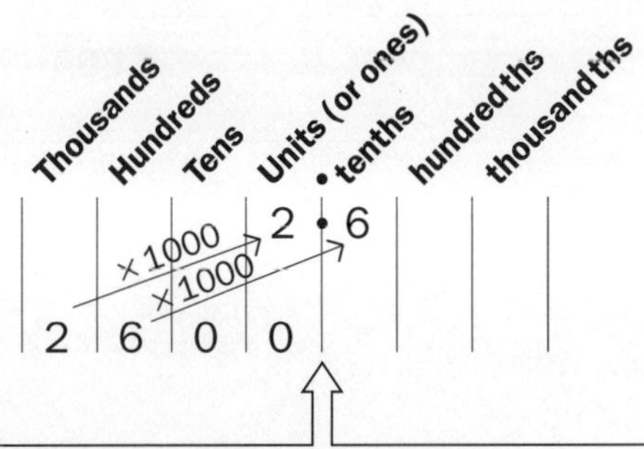

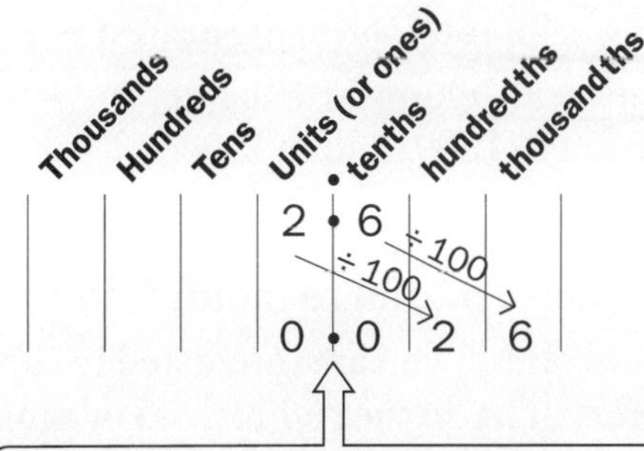

This diagram shows what happens when you multiply 2·6 by 1000.
Each figure moves three columns to the left.
The zeros show that there are no units or tens.

This diagram shows what happens when you divide 2·6 by 100.
Each figure moves two columns to the right.
The zero after the decimal point shows that there are no tenths.

You need to be able to do multiplications and divisions like these in your head.

$12 \times 100 = 1200$ $\quad 680 \div 10 = 68$ $\quad 50 \times 20 = 1000$ $\quad 200 \div 4 = 50$

Check that you agree with the answers.

8 Work out each of these in your head.

(a) 300×3 (b) 4×50 (c) 1000×4 (d) 30×40 (e) 700×20
(f) $500 \div 5$ (g) $800 \div 10$ (h) $100 \div 20$ (i) $2000 \div 4$ (j) $3000 \div 6$
(k) 66×100 (l) $530 \div 10$ (m) 72×1000 (n) 81×100 (o) $1450 \div 10$

9 Write down the answer to each of these. Use a calculator if you wish.

(a) $5{\cdot}4 \times 10$ (b) $8{\cdot}1 \times 100$ (c) $0{\cdot}8 \times 10$ (d) $1{\cdot}02 \times 10$ (e) $0{\cdot}07 \times 100$
(f) $7{\cdot}2 \div 10$ (g) $757 \div 100$ (h) $88 \div 100$ (i) $810 \div 1000$ (j) $3{\cdot}3 \div 100$

10 Write down the missing symbol (× or ÷) which makes each of these true:

(a) $150 \square 10 = 1500$ (b) $30 \square 30 = 1$ (c) $9300 \square 100 = 93$ (d) $37 \square 100 = 3700$

11 An exercise book is 0·6 cm thick.
How high will a pile of 100 similar exercise books be?

12 A stack of 10 sheets of plastic has a thickness of 1·2 cm.
What is the thickness of each sheet?

13 Kieron thinks that **3·2 × 10 = 3·20**.
How would you explain to Kieron why his answer must be wrong?
It is not enough just to give the correct answer.

MEG (SMP)

Answers and hints ► page 100

Rounding

Rounding to the nearest ten, hundred, thousand, ...

Example Round 1568 to the nearest hundred.

1568 is between 1500 and 1600, but
1568 is nearer to 1600 than to 1500.
So 1568 is rounded up to 1600 (the nearest hundred).

There is a simple **rule for rounding**:

- Find the column you are interested in (units, tens, hundreds, ...)
- If the next digit on the right is: 5 or more, round up
  4 or less, round down

Examples

48 125 rounded to the nearest 1000 is 48 000. ⇐ *The digit after the thousands is 1, so round down.*

982 rounded to the nearest 100 is 1000. ⇐ *The digit after the hundreds is 8, so round up.*

655 rounded to the nearest 10 is 660. ⇐ *655 is exactly halfway between 650 and 660. The usual rule is to round up.*

1 Round each of these numbers to the nearest hundred.
(a) 4760 (b) 72 340 (c) 3025 (d) 4350 (e) 979 (f) 3333

2 Round 25 377 (a) to the nearest ten (b) to the nearest thousand

3 Round 4 810 737 (a) to the nearest million (b) to the nearest thousand

4 The table gives the population of the two largest cities in Wales as given in the 1990 AA handbook.

Swansea	282 610
Cardiff	279 500

(a) What is the population of Swansea to the nearest thousand?
(b) To what accuracy has the population of Cardiff probably been rounded?

5 *Sportstime* has a phone-in competition.

'Guess how many people will attend the football match at Southampton today.'

Some of the entries are: 14 284 14 380 14 231 14 314
14 394 14 376 14 351 14 280

The club reports that the attendance is 14 300 to the nearest hundred.
Which of the competition entries cannot be winners?

6 (a) Nicky won £5 868 631 in the lottery.
To what accuracy is this headline in a local newspaper?

LOCAL WOMAN WINS £6 000 000 IN LOTTERY

(b) The local football match had a record gate of 12 788.
The local paper rounds this figure to the nearest thousand.
Copy and complete this headline: RECORD GATE OF ....

Rounding to a number of decimal places

Example Round 1·24 to 1 decimal place.

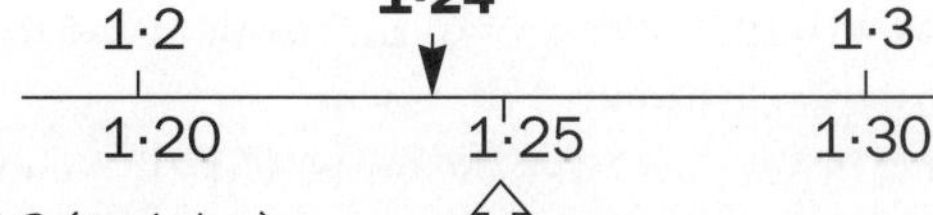

1·24 is nearer to 1·2 than to 1·3.
So 1·24 is rounded down to 1·2.

We write 1·24 = 1·2 (to 1 decimal place) or 1·24 = 1·2 (to 1 d.p.)

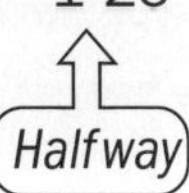

There is a simple rule for rounding decimals:

- Find the column you are interested in (1st, 2nd, 3rd,... decimal place)
- If the next digit on the right is: 5 or more, round up
  4 or less, round down

Example Round 7·3278149 to 2 decimal places.

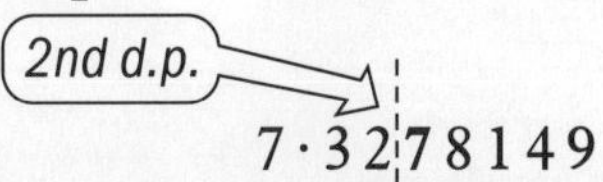

7·32¦78149

The number in the column to the right of the 2nd decimal place is 7.
So round up to 7·33.

Rough estimates ► page 8

Rounding to the nearest penny

£5·1428 = £5·14 (to the nearest penny) ⇐ *The 3rd digit after the decimal point is 2, so round down.*

7 Round these to 2 decimal places.
(a) 2·4826 (b) 9·308 (c) 5·0078 (d) 4·279 (e) 0·3965 (f) 0·7047

8 Round these to 1 decimal place.
(a) 2·1946 (b) 7·628 (c) 0·068 (d) 11·9082 (e) 4·9616 (f) 6·0192

9 Round the number on this calculator display to
(a) 1 d.p. (b) 2 d.p. (c) 3 d.p. (d) 4 d.p.

5.8902883

10 Round off these amounts to the nearest penny.
(a) £7·523 (b) £9·628 (c) £5·709 (d) £4·0791 (e) £10·489 (f) £0·796

11 Use a calculator to work out these.
Give your answer to the number of decimal places shown.

Using a calculator ► page 10

(a) 26·19 – 4·85 (1 d.p.) (b) 2·35 × 7·91 (2 d.p.) (c) 2800 ÷ 46 (2 d.p.)
(d) 14 ÷ 2·7 (1 d.p.) (e) 0·481 + 6·985 (2 d.p.) (f) 0·576 ÷ 9·9 (3 d.p.)

12 Ella buys this pack of six batteries for £3·50.
How much does one battery cost to the nearest penny?

MEG/ULEAC (SMP)

13 Petrol costs 63·9p per litre.
Darren buys 46 litres of petrol.
How much does it cost him to the nearest penny?

Answers and hints ► page 100

Pencil and paper calculations

You will have your own methods for doing addition, subtraction, multiplication and division.
It is important to show your working so you can gain marks for method even if you make a mistake in the working.

Checking

You can check a subtraction by adding.

$$\begin{array}{r} 184 \\ -115 \\ \hline 69 \end{array} \qquad 115 + 69 = 184$$

You can check a division by multiplying.

$$8\overline{)34^{2}4} = 4\ 3 \qquad \begin{array}{r} 43 \\ \times\ 8 \\ \hline 344 \end{array}$$

Do not use a calculator for any of these questions. Show all your working.

1 (a) $874 + 58$ (b) $23 + 752$ (c) $563 - 231$ (d) $870 - 358$ (e) $901 - 722$

2 (a) 68×23 (b) 369×37 (c) $992 \div 8$ (d) $495 \div 15$ (e) $782 \div 23$

3 A coach can hold 53 passengers.
28 passengers are seated in the coach.
How many empty seats are there?

4 Tom and Richard are keen cricketers.
In his first three innings Tom scores 54, 38 and 63 runs.
In his first three innings Richard scores 19, 44 and 76 runs.
How many more runs than Richard does Tom score?

5 A carpenter is hanging the doors in a new house.
There are 9 doors and he needs 2 hinges and 16 screws for each door.
(a) How many hinges does he need?
(b) How many screws does he need?

6 96 children are going to have dinner.
(a) How many tables like this one will they need?
(b) How many benches will they need?

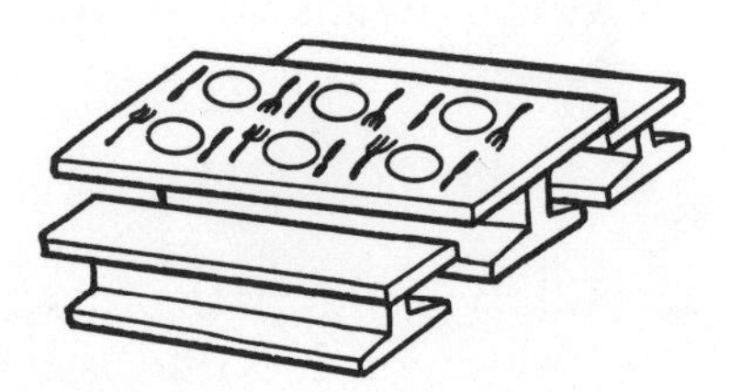

7 Three figures are missing in this subtraction.
Copy and complete the calculation to work out the missing numbers.

$$\begin{array}{r} \square\ 5\ 8 \\ -\ 2\ \square\ 3 \\ \hline 4\ 8\ \square \end{array}$$

8 (a) Sally buys seven custard slices costing 34p each.
She pays with a £5 note. How much change should she be given?

(b) Simon buys 8 CDs for £98. All the CDs are the same price.
What is the cost of each CD?

WJEC

9 A crate can hold 24 bottles.
How many crates are needed to hold 840 bottles?

10 Crisps cost 32p per packet.
Work out the cost of 36 packets in pounds.

11 Alex has a holiday job for 12 weeks. She takes home £65 per week.

(a) How much will she earn altogether?

(b) She budgets to give her mother £15 per week and
to spend £10 per week on herself.
How much should she be able to save in 12 weeks?

12 368 chairs are arranged in rows of 16.
How many rows are there?

13 (a) Emma thinks of a number and multiplies it by 2. Her answer is 58.
What number is Emma thinking of?

(b) Aziz thinks of a number and adds 23 to it. His answer is 71.
What number is Aziz thinking of?

14 A school has a total of 912 pupils; 474 of them are girls.
How many more girls than boys are there?

15 John is on holiday in France.
He buys some bottles of wine which cost 14 francs each.
He spends a total of 252 francs.

(a) How many bottles does he buy?

(b) John pays with three 100 franc notes.
How much change does he get?

MEG (SMP)

16 Rashid is on holiday in Greece.
He wants to change £50 into drachmas.
The bank pays 420 drachmas for every £1,
but charges a fee of 800 drachmas for changing the money.

How many drachmas will Rashid get?

Answers and hints ► page 101

Rough estimates

Example

Decide whether you can buy five T-shirts at £3·95 each with a £20 note.

Each T-shirt costs just under £4.
Five T-shirts will cost just under $£4 \times 5 = £20$.
So you can buy the five T-shirts.

Rounding ► page 4

1 Pete has a £25 M & S voucher for his birthday.
He lists the clothes he would like to buy.
Write down a rough calculation he could do in his head to see if he could afford them.

Leather belt	£11·99
Pair of socks	£3·90
T-shirt	£8·50

2 You want an approximate answer for 596×68.

600×60	500×70
600×70	500×60

(a) Which of the approximations in the box is closest?

(b) Explain whether this approximate answer is too large or too small.

3 Nylon rope costs £2·89 per metre.
Estimate the cost of 210 metres of the rope.

4 (a) Becky wants to buy 2 tapes and a single.
Work out in your head a rough answer for the total cost.
Show how you worked out your **rough** answer.

BARGAINS

```
Tapes .........£3·99
Singles .......£1·99
CDs ...£5·99 to £7·99
```

(b) Robert buys 6 CDs. They cost £7·99 each.
Which amount is closest to the total cost of his CDs?

£40 **£45** **£50** **£55**

Explain how you chose your answer.

MEG (SMP)

5 Sushma is paid £20 for babysitting for 4 hours 45 minutes.
Roughly how much per hour is this?

6 Mr Khan wants to carpet a room. He needs $49\,m^2$ of carpet.

(a) The carpet costs £9·85 per square metre.
He wants to estimate the cost of $49\,m^2$ of this carpet.
Write down a calculation he could do in his head.

(b) Explain how you know whether his estimate is bigger or smaller than the exact cost. MEG (SMP)

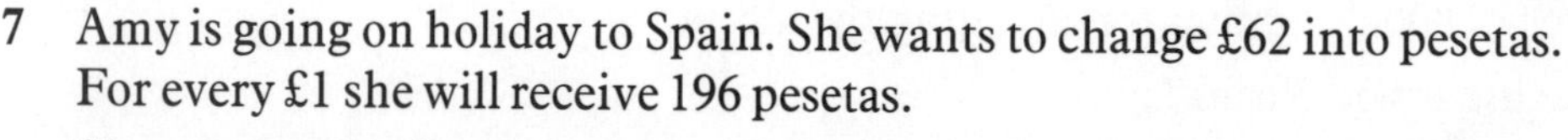

7 Amy is going on holiday to Spain. She wants to change £62 into pesetas.
For every £1 she will receive 196 pesetas.

She wants to **estimate** how many pesetas she will receive.
Write down a calculation she could do in her head, and the answer.

MEG (SMP)

Answers and hints ► page 102

Negative numbers

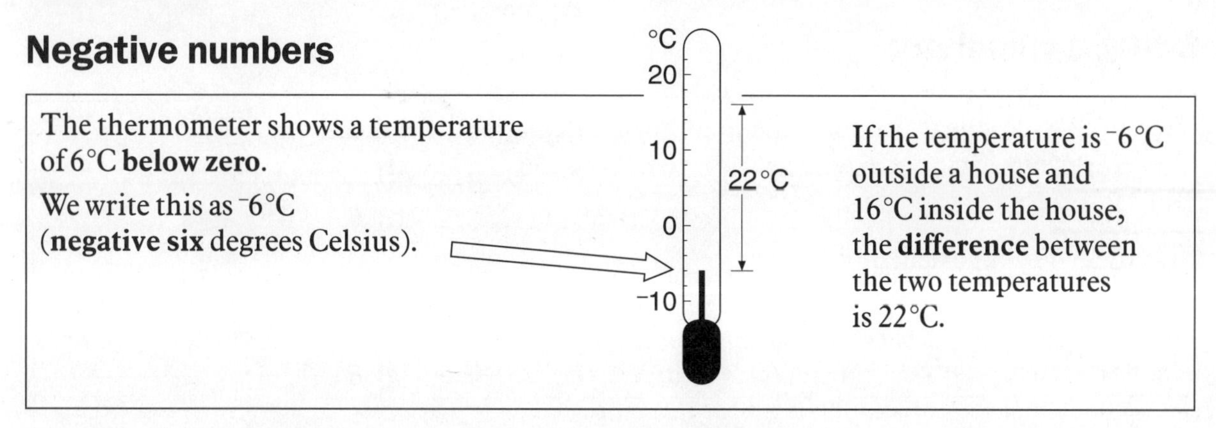

The thermometer shows a temperature of 6°C **below zero**.
We write this as $^-6$°C (**negative six** degrees Celsius).

If the temperature is $^-6$°C outside a house and 16°C inside the house, the **difference** between the two temperatures is 22°C.

1 Write these temperatures in order, starting with the lowest.

$^-7$°C 3°C 0°C $^-2$°C 4°C

2 Sharon and Tessa are camping.

(a) At 7 a.m. one morning the temperature outside the tent was $^-4$°C.
The temperature inside the tent was 3°C higher.
What was the temperature inside the tent?

(b) The temperature at 7 a.m. was $^-4$°C.
At midday it was 5°C.
How many degrees had it gone up?

(c) On the first three mornings of their trip the temperatures were

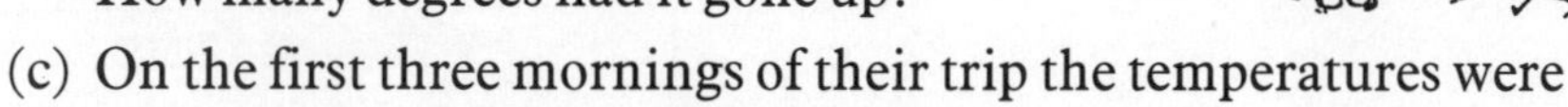

$^-1$°C $^-4$°C $^-2$°C.

Write these in order, starting with the coldest.

MEG/ULEAC (SMP)

3 The table shows the temperature round Britain one day in winter.

(a) Which city had the lowest temperature?

(b) Which city had the highest temperature?

(c) The temperature in Dublin was two degrees lower than that in Cardiff.
Which city had the same temperature as Dublin?

NICCEA

City	Temperature
Aberdeen	$^-3$°C
Belfast	$^-1$°C
Cardiff	0°C
Glasgow	$^-2$°C
London	2°C
Newcastle	1°C
Plymouth	3°C
Wolverhampton	$^-1$°C

4 Satellites in space have to withstand large changes in temperature.
On the sunlit side the temperature of the metal is 150°C.
On the shaded side it is $^-120$°C.

What is the difference between these two temperatures?

MEG/ULEAC (SMP)

Answers and hints ► page 102

Using a calculator

You need to be able to use your calculator to do calculations like these.

(a) $7 \times (2+5)$ (b) $(9+3) \div 12$ (c) $(28 \div 4) - 3$ (d) $12 + (4 \times 3)$

(e) $\frac{10-2}{4}$ (f) $4 + \frac{10}{5}$ (g) $3^2 - 1$ (h) $6 + \sqrt{4}$

Do them on your calculator.
Then check your answers by working them out in your head.

Rounding ► page 4
Rough estimates ► page 8

If you got any wrong, find out where you went wrong.
The answers are: (a) 49 (b) 1 (c) 4 (d) 24 (e) 2 (f) 6 (g) 8 (h) 8

1 Use a calculator to work out the following.

(a) $\frac{10{\cdot}32 + 6{\cdot}69}{2{\cdot}7}$ (b) $\frac{60{\cdot}24 - 29{\cdot}74}{50}$ (c) $\frac{1{\cdot}2^2}{1{\cdot}8}$

(d) $4{\cdot}2 \times (3{\cdot}5 + 2{\cdot}85)$ (e) $(20{\cdot}6 \times 1{\cdot}2) - 3{\cdot}6$ (f) $7{\cdot}8 + \sqrt{3{\cdot}61}$

Check your answers by making rough estimates.

2 Fabric is £4·99 a metre.
Sabrina buys 2·1 metres of fabric.

(a) How much does Sabrina have to pay?

(b) Write down a calculation you could do in your head to check your answer.

MEG (SMP)

3 Jacinta walked 2·52 kilometres in the morning, 4·19 kilometres in the afternoon and 1·87 kilometres in the evening.

(a) How far did she walk altogether?

(b) How much further did she walk in the morning than in the evening?

NICCEA

4 Find the value of each of these.
Give your answers correct to 1 decimal place.

(a) $2{\cdot}15^2 + 6{\cdot}24$ (b) $\sqrt{75}$ (c) $1{\cdot}8^3$

5 Jalal weighs 86·4 kg. His sister weighs 62·9 kg.
How much heavier is Jalal than his sister?

WJEC

6 Hailey spends £25·00 on petrol which costs 69·5p a litre.
How many litres does she get, to the nearest litre?

7 Laura buys 3 cans of coke which cost £0·45 each and a bottle of lemonade which costs £1·12.
She pays for the drinks with a £5 note.
What change should she be given?

8 This table gives the number of bookings of different types of holiday taken with a travel company in 1994.

City Breaks	762
Car Tours	214
Camping Holidays	103
Log Cabins	84
Theme Holidays	165
Apartments	59
Hotels	2467

(a) Work out the total number of bookings in 1994.

(b) Write your total to the nearest 100.

MEG (SMP)

9 Chris hires a van for two days and drives a total of 450 miles. How much will it cost him to hire the van altogether?

VAN HIRE
£42 per day
Plus £0·18 per mile

10 A firm pays £4·00 for delivering 200 leaflets.

(a) How much is this per leaflet?

(b) Irene was paid £7·00. How many leaflets did she deliver?

MEG (SMP)

11

ASSISTANT COOK **£4·22 per hour**
To work 28 hours per week at the Chuter Ede Primary School, Wolfit Avenue.

Work out the assistant cook's full weekly wage.

ULEAC

12 Jo and her three friends go to a pop concert as a birthday treat. The tickets cost £18·75 each and the bus fares £2·30 each. How much does the treat cost altogether?

13 A three-piece suite costs £699 cash. Jahal buys it on hire purchase. He pays £100 deposit and 36 monthly instalments of £24·50. How much less would he pay if he paid cash?

WJEC

14 Mrs Kahn works a basic 36-hour week at an hourly rate of £6·26. One week she also does 3 hours overtime at time and a half and 4 hours overtime at double time. What is her total pay for the week?

Answers and hints ► page 102

Leftovers

In division problems that don't work out exactly, you often need to decide whether to round up or down.

If you were answering the question
'How many 6-egg boxes do you need to pack 31 eggs?'
you would need to **round up**.

The answer is 6. You need a box for the one egg left over.

If you were answering the question
'How many teams of six can you enter for a cross-country race from 31 athletes?'
you would need to **round down**.

The answer is 5. One athlete is left out.

1 Party invitations are sold in packs of six.
Latoya needs enough invitations for 75 people.
How many packs will she need to buy?

2 Jenny goes into a Post Office with a £5 note and buys as many 19p stamps as possible.
(a) How many stamps does she buy?
(b) How much change does she get?
MEG/ULEAC (SMP)

3 A cable car taking people to the top of a mountain can hold a maximum of 17 people.
What is the least number of times the cable car will have to go up the mountain in order to take 550 people to the top?
WJEC

4 Alan is buying plastic glasses for a party.
They are sold only in packs of 12 glasses.
Each pack costs £1·89.
Alan needs 40 glasses.
(a) How many packs must he buy?
(b) How much does he pay?

MEG (SMP)

5 A stack of 13 identical shoe boxes is 163·8 cm high.
(a) Calculate the thickness of each shoe box.
(b) How many of the shoe boxes could be stacked one on top of the other between two shelves which are 100 cm apart?

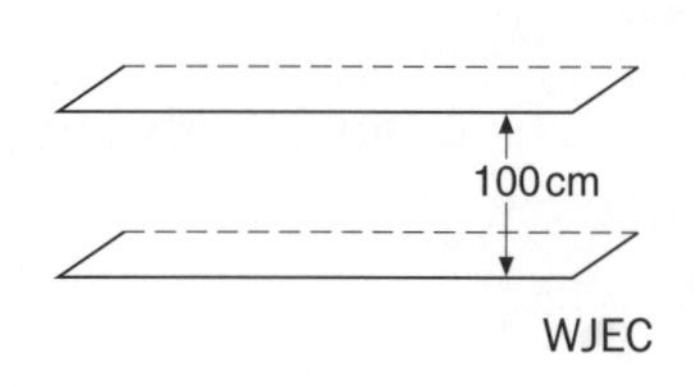

WJEC

6 209 people have bought tickets for a school concert.
Seats in the school hall are to be set out in complete rows.
Each row is to have 24 seats.
(a) What is the least number of complete rows needed?
(b) How many spare seats will there then be?
MEG/ULEAC (SMP)

Answers and hints ► page 104

Trial and improvement

Example

?cm

?cm

A square has an area of 22 square centimetres.
Find the length of its sides to 1 decimal place without using a square root key.

A square of side 4 cm has an area of 16 cm^2. *Too small*
A square of side 5 cm has an area of 25 cm^2. *Too large*

So the length of a side must lie between 4 cm and 5 cm.

You can use a table like this.
Try different values for the length, starting with, say, 4·5 cm.

Keep on trying and improving until you have an answer which is accurate enough.

Length tried	Working	
4·5 cm	4·5 × 4·5 = 20·25	Too small
4·6 cm	4·6 × 4·6 = 21·16	Too small
4·7 cm	4·7 × 4·7 = 22·09	Just too large

The length of the sides of the square must be 4·7 cm (to 1 d.p.). *4·7 is much nearer than 4·6.*

This method is called **trial and improvement**.

Rounding to decimal places ► page 5

1 Mike is designing some publicity cards.
The area of each card must be 40 cm^2 and the card must be square.
He tries to find the width of the card by trial and improvement.
His first try was width 5 cm. 5 × 5 = 25 Too small
His second try was width 7 cm. 7 × 7 = 49 Too large
He wrote the results in a table like this.
Copy the table.

Width tried	Working	
5 cm	5 × 5 = 25	Too small
7 cm	7 × 7 = 49	Too large
6 cm	6 × 6 =	

(a) Try width 6 cm.
Is 6 too small or too large?
Complete the table for 6 cm.

(b) Try at least three other values for the width to find the area correct to 1 decimal place.
Record each value you try in the table.

(c) Write down the value you would choose for the width of the card.

MEG (SMP)

2 Simon is trying to find the square root of 61 without using the square root key on his calculator.
He tries 7 and gets 7 × 7 = 49 which is too small.
Then he tries 8 and gets 8 × 8 = 64 which is too large.
Make a table like the ones above to find the square root of 61 correct to 1 d.p.

3 A sheep pen is 2 metres longer than it is wide.

(a) Calculate the area of a pen which is (i) 4 m wide (ii) 5 m wide.

(b) Use trial and improvement to find the width of a pen which has an area of 32 m^2.
Give your answer in metres correct to 1 decimal place.

Answers and hints ► page 104

Fractions

Equivalent fractions

The number on the bottom of a fraction tells you the number of **equal** parts.
The number on the top tells you how many parts you are counting.

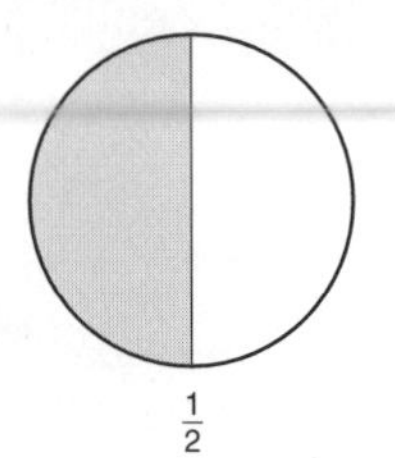
$\frac{1}{2}$

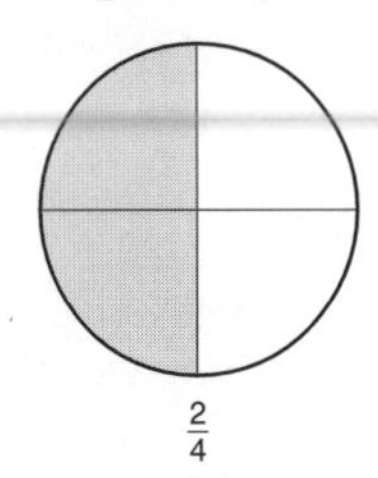
$\frac{2}{4}$

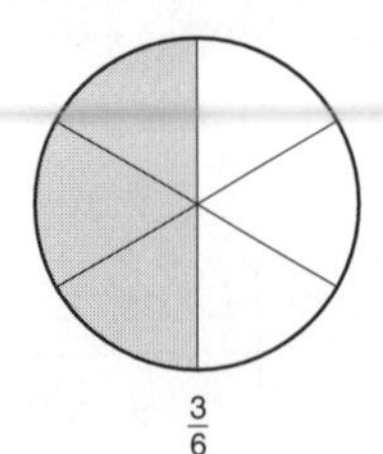
$\frac{3}{6}$

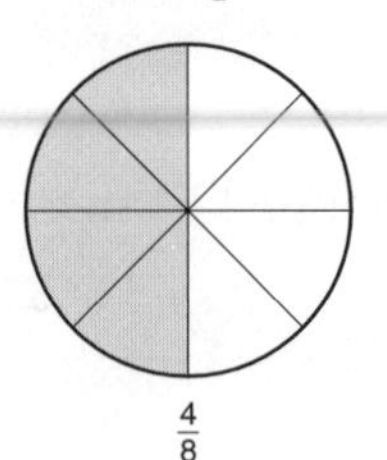
$\frac{4}{8}$

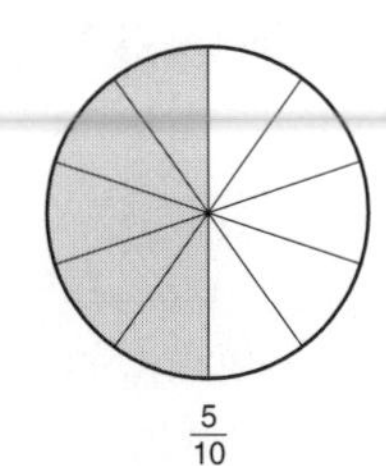
$\frac{5}{10}$

$\frac{1}{2} = \frac{2}{4} = \frac{3}{6} = \frac{4}{8} = \frac{5}{10}$ are called **equivalent fractions.**

$\frac{1}{2}$ is the **simplest form** of the fractions.

Addition and subtraction

To add and subtract fractions make sure they have the same bottom number.

Examples

Work out: (a) $\frac{3}{8} + \frac{1}{4}$ (b) $\frac{15}{16} - \frac{1}{2}$ (c) $2\frac{3}{4} - \frac{1}{2}$

(a) $\frac{3}{8} + \frac{1}{4}$ is the same as $\frac{3}{8} + \frac{2}{8} = \frac{5}{8}$ ⇐ *It is easy to add the fractions here.*

(b) $\frac{15}{16} - \frac{1}{2}$ is the same as $\frac{15}{16} - \frac{8}{16} = \frac{7}{16}$

(c) $2\frac{3}{4} - \frac{1}{2} = 2 + \frac{3}{4} - \frac{1}{2}$ which is the same as $2 + \frac{3}{4} - \frac{2}{4} = 2 + \frac{1}{4} = 2\frac{1}{4}$. ⇐ This is called a **mixed number.**

Finding a fraction of a quantity

Examples

To work out $\frac{1}{4}$ of something, you just divide by 4. ⟹ $\frac{1}{4}$ of £20 is £20 ÷ 4 = £5.

To work out $\frac{3}{5}$ of £20, you first work out $\frac{1}{5}$ of £20. ⟹ £20 ÷ 5 = £4

$\frac{3}{5}$ is 3 times $\frac{1}{5}$, so $\frac{3}{5}$ of £20 is £4 × 3 = £12.

Fractions, decimals and percentages ► page 16

1 Write down three fractions that are equal to $\frac{1}{3}$.

2 (a) Copy this shape on squared paper and shade in $\frac{2}{3}$ of it.

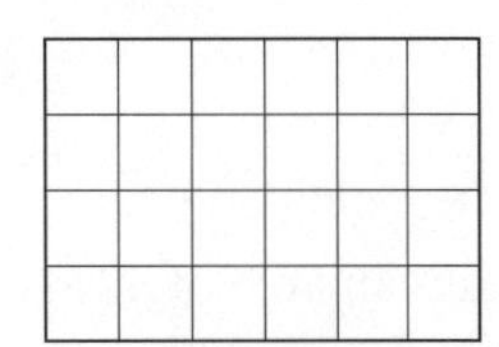

(b) (i) What fraction of this shape is shaded? Give your fraction in its simplest form.

(ii) What fraction is not shaded?

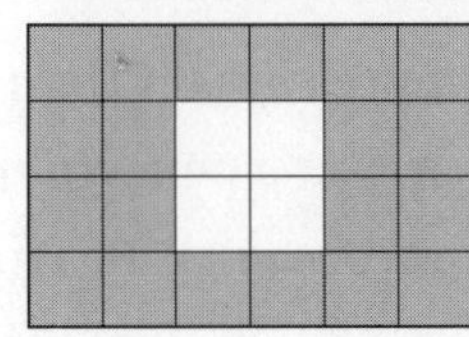

MEG (SMP)

3 Work out these.

(a) $\frac{1}{8} + \frac{3}{4}$ (b) $\frac{7}{8} - \frac{3}{16}$ (c) $1\frac{1}{2} + \frac{3}{8}$ (d) $\frac{1}{2} - \frac{3}{16}$ (e) $2\frac{3}{8} + 1\frac{1}{4}$

4 Which number matches each arrow?

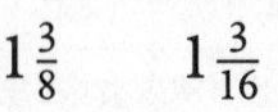
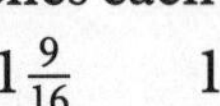

$1\frac{3}{8}$ $1\frac{3}{16}$ $1\frac{9}{16}$ $1\frac{3}{4}$

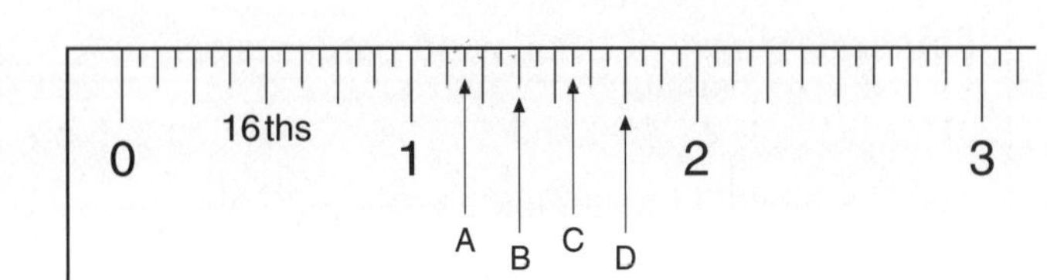

5 Work out each of these.

(a) $\frac{1}{2}$ of £5 (b) $\frac{1}{3}$ of 24 cm (c) $\frac{2}{5}$ of £35 (d) $\frac{3}{4}$ of 500 g

6 In a raffle Lee won a box of fruit.
It contained: $1\frac{1}{4}$ lb of bananas, 1 lb of apples,
$\frac{3}{4}$ lb of oranges, $\frac{1}{2}$ lb of grapes.

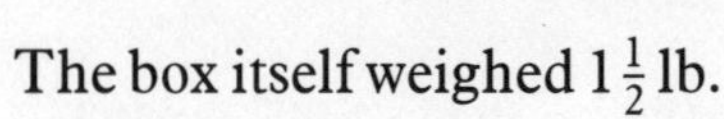

The box itself weighed $1\frac{1}{2}$ lb.

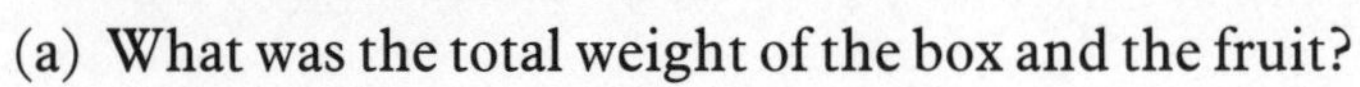

(a) What was the total weight of the box and the fruit?

Lee and his friends ate half the apples and half the grapes.

(b) (i) What weight of fruit did they eat?
(ii) What was the total weight left for Lee to carry home?

MEG/ULEAC (SMP)

7 40 pupils take part in a sponsored swim.
15 pupils swim more than 1 mile.

What fraction of pupils swim more than 1 mile?
Give your answer in its simplest form.

8 A bar of chocolate has 20 pieces.

(a) Gurdeep breaks off five pieces.
What fraction of the bar is this?

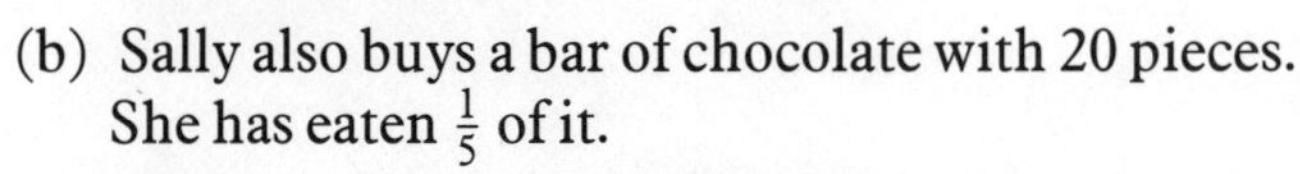

(b) Sally also buys a bar of chocolate with 20 pieces.
She has eaten $\frac{1}{5}$ of it.
How many pieces has she eaten?

SEG

9 Write these fractions in their simplest form.

(a) $\frac{2}{16}$ (b) $\frac{12}{16}$ (c) $2\frac{8}{16}$

10 When a pendant is carved from a block of wood, $\frac{4}{9}$ of the wood is cut away.
The original block of wood weighs 225 grams.
What weight of wood is cut away?

MEG

Answers and hints ► page 105

Fractions, decimals and percentages

Converting between fractions, decimals and percentages

This scale shows decimals and percentages.

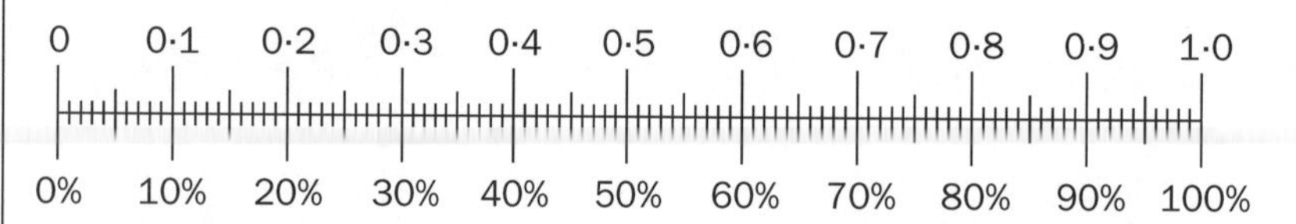

10% is the same as 0·1.
56% is the same as 0·56.
5% is the same as 0·05.

You can see that 35% = 0·35, but you can also work it out like this.

$$35\% = \frac{35}{100} = 35 \div 100 = 0{\cdot}35$$

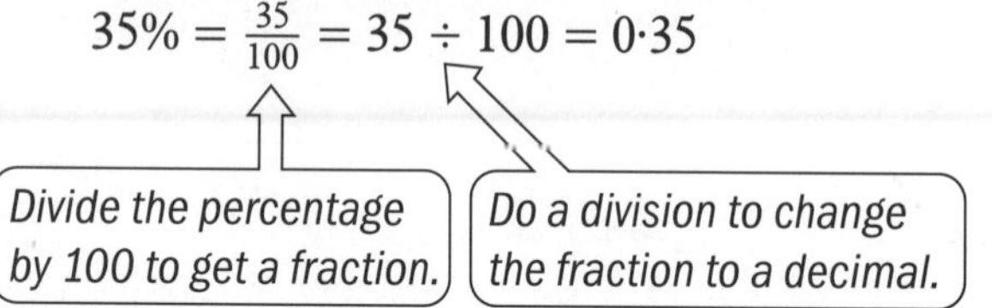

1% of something means $\frac{1}{100}$ of it.
35% of something means $\frac{35}{100}$ of it.

To change a decimal to a percentage you multiply it by 100: $0{\cdot}45 = 0{\cdot}45 \times 100\% = 45\%$

Percentages ► page 18

1 Write these percentages as fractions.

(a) 25% (b) 30% (c) 35% (d) 68% (e) 21% (f) 4%

2 Use a calculator to change these fractions to decimals.

(a) $\frac{1}{2}$ (b) $\frac{3}{4}$ (c) $\frac{7}{10}$ (d) $\frac{3}{8}$ (e) $\frac{17}{100}$ (f) $\frac{9}{100}$

3 Write these decimals as percentages.

(a) 0·8 (b) 0·71 (c) 0·32 (d) 0·07 (e) 0·025 (f) 0·375

4 Write these decimals and percentages as fractions.

(a) 30% (b) 0·7 (c) 25% (d) 0·21 (e) 0·03 (f) 74%

5 Write these fractions as percentages.

(a) $\frac{1}{4}$ (b) $\frac{1}{5}$ (c) $\frac{3}{4}$ (d) $\frac{1}{8}$ (e) $\frac{36}{100}$ (f) $\frac{9}{100}$

6 Seven-eighths of this rectangle is shaded.

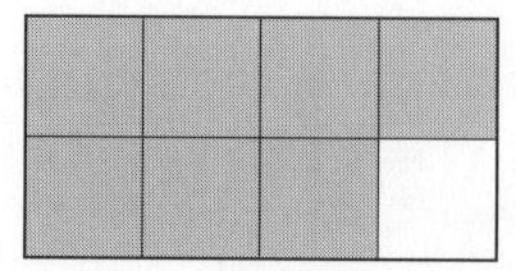

(a) Write $\frac{7}{8}$ as a decimal.

(b) Write $\frac{7}{8}$ as a percentage.

MEG (SMP)

7 Fran planted 15 bulbs in a bowl. 12 grew into flowers.

(a) Write $\frac{12}{15}$ as a decimal.

(b) What percentage of the bulbs grew into flowers?

MEG (SMP)

8 Put these in order, from the smallest to the largest: 0·3 $\frac{1}{2}$ 20% $\frac{1}{4}$ $\frac{1}{3}$ 24%

Calculating a percentage of a quantity

Remember:

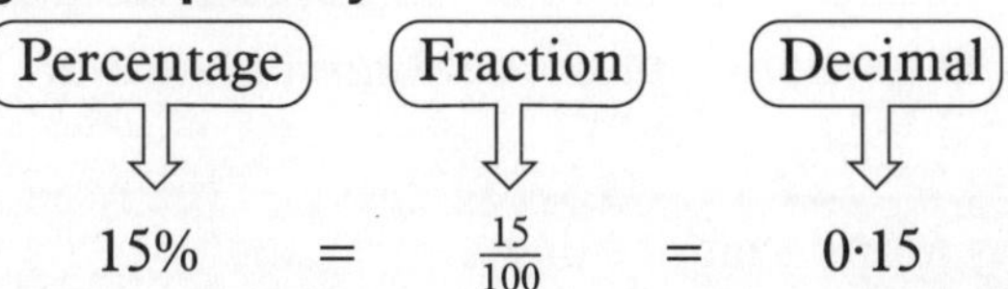

$$15\% = \frac{15}{100} = 0{\cdot}15$$

Example

The normal pricc of a computer is £350.
This price is reduced by 15% in a sale.
What is the reduction?

Change the percentage to a decimal. ⇨ 15% = 0·15

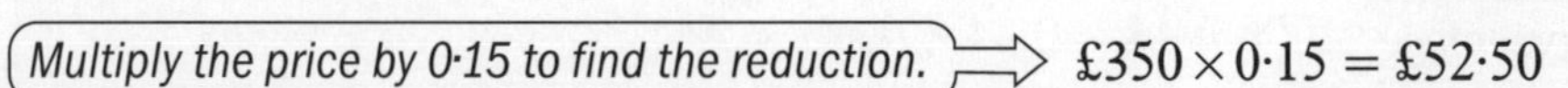

Multiply the price by 0·15 to find the reduction. ⇨ £350 × 0·15 = £52·50

The reduction (or discount) is £52·50.

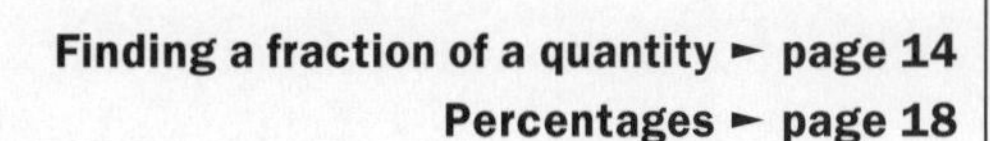

Finding a fraction of a quantity ► page 14
Percentages ► page 18

9 A charity has an income of £350 000.
It spends 12% on administration. How much is this?

10 A school raises £1360 from a sponsored swim.
$\frac{3}{8}$ of the money is given to Oxfam. How much is this?

11 A car manufacturer increases its price by 4%.
How much extra would you have to pay on a car costing £11 400?

12 In a survey, 500 people were questioned about things they recycled.

(a) 25% of the people said they recycled paper.
How many people is this?

(b) $\frac{7}{10}$ of the people said they recycled bottles.
How many people is this?

ULEAC (KMP)

13 The number of people living in cities is increasing.
In 1975 about two people in every ten in the world lived in a city.

(a) What percentage of the world lived in cities in 1975?

(b) By the year 2000 the population of the world will be about 9000 million.
About 60% of these people will live in cities.
About how many millions will live in cities?

MEG/ULEAC (SMP)

14 An electrician charges £27·50 for a service call.
VAT is added at a rate of 17·5%.

How much VAT would you pay?
Give your answer to the nearest penny.

Answers and hints ► page 105

Percentages

Expressing one quantity as a fraction or percentage of another

Example

Class 8A did a survey. They found that one day
330 out of 750 pupils came to school by car.
What percentage came to school by car?

Write the information as a fraction. ⟹ $\frac{330}{750}$

Change the fraction to a decimal. ⟹ $\frac{330}{750} = 330 \div 750 = 0{\cdot}44$

Change the decimal to a percentage. ⟹ $0{\cdot}44 = 0{\cdot}44 \times 100\% = 44\%$

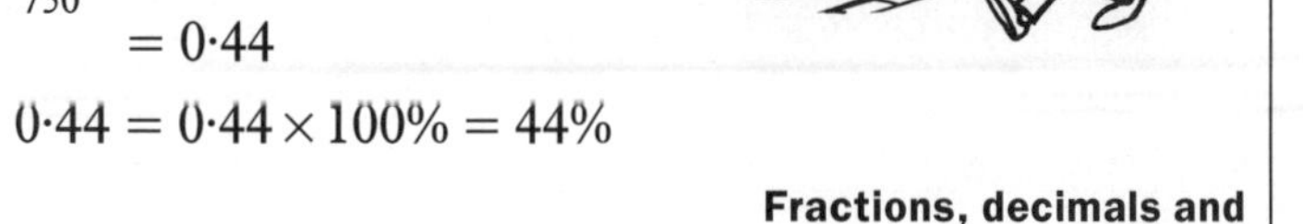

44% of pupils came to school by car.

Fractions, decimals and percentages ► page 16

Finding a value after a percentage increase or decrease

Example

A builder charges £2400 for some work. He has to add 17·5% VAT to his bill.
What will the total bill come to?

Change the percentage to a decimal. ⟹ $17{\cdot}5\% = 0{\cdot}175$

Multiply the price by 0·175 to find the VAT. ⟹ $£2400 \times 0{\cdot}175 = £420$

Add the increase to the original price. ⟹ $£2400 + £420 = £2820$

The total bill will be £2820.

1 The normal price of this video player is £350.
There is 15% off this price in the sale.

(a) Write 15% as a decimal.

(b) How much is 15% of £350?

(c) How much will the video player cost in the sale?

MEG (SMP)

2 Sui earns £135 per week. She gets a 7% pay rise.

(a) How much is her pay rise?

(b) How much does she get each week after her pay rise?

WJEC

3 A block of metal weighs 800 grams.
300 grams is removed on a lathe.
What percentage is removed?

4 (a) Maninder measures the height of a flower as 60 cm.
The next day it is 15% taller.
What is its new height?

(b) Another plant grows from 1·6 m to 1·8 m.
Calculate the percentage change in height.

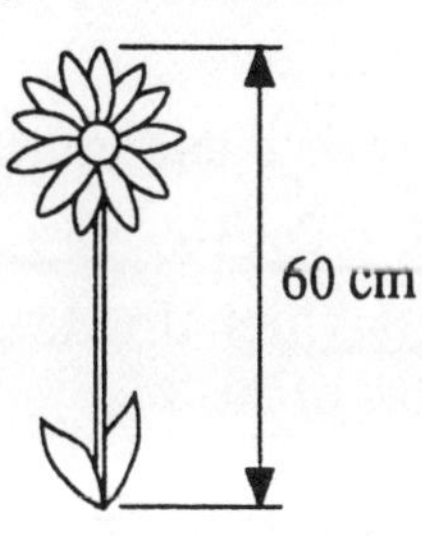

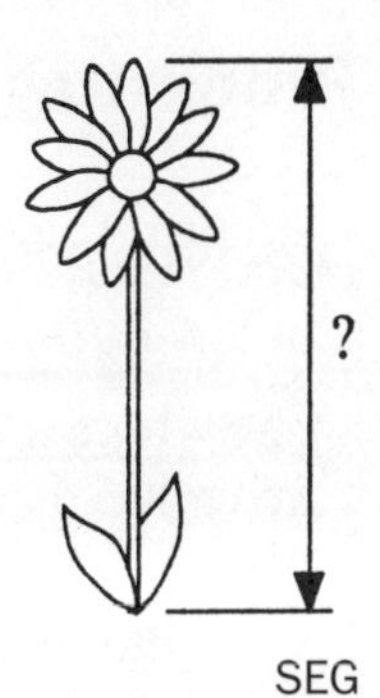

SEG

5 The percentage of people unemployed in Great Britain was highest in 1933.
Out of the 13 million people available for work 3 million were unemployed.

What percentage were unemployed?
Give your answer to the nearest whole number.

6 A company employs 348 women and 252 men.
What percentage of the workforce are women?

7 A 340 g jar of jam contains 35% fruit.
What weight of fruit does the jar contain?

8 In a sale all goods are reduced by 18%.
What will be the price of this cassette player in the sale?

9 At the beginning of a school year Hollybrook School had 1100 pupils.
By the end of the year there were 77 fewer pupils.

What is the percentage decrease in the number of pupils?

10 Gareth and Vicky worked in different departments of *Star Enterprises*.

Gareth earned £645 a month and was given a pay rise of 7%.
Vicky earned £670 a month and was given a pay rise of 5%.

Who had the higher monthly pay after the increases?

11 Andrew wants to buy a midi-system.
Bargain Stores and *Best Buys* usually sell it for £320.
Bargain Stores has it on special offer at £289·99.
Best Buys are having a sale and reducing all prices by 15%.

At which store is the midi-system cheaper, and by how much?

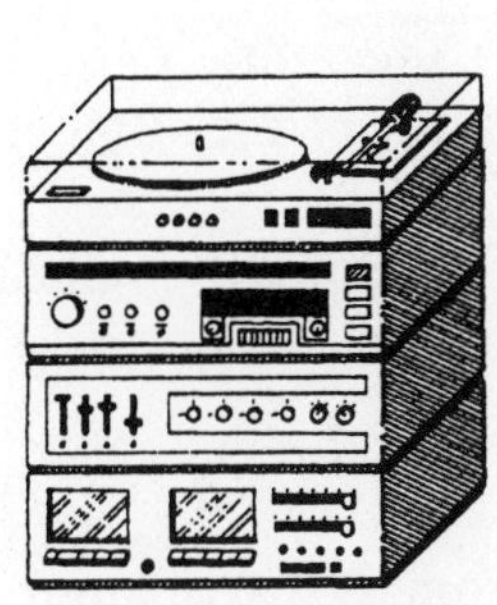

Answers and hints ► page 106

Properties of numbers

Factors, prime numbers and multiples

$20 = 1 \times 20$ or $20 = 2 \times 10$ or $20 = 4 \times 5$

The whole numbers which divide exactly into 20 are called the **factors** of 20.

20 has six factors.
They are 1, 2, 4, 5, 10 and 20.

A number which has only two factors, itself and 1, is called a **prime** number.

2, 3, 5 and 7 are all prime numbers, but 1 is not. (1 has only one factor.)

Any number which is not a prime number can be split into the product of prime factors. For example:

$$18 = 2 \times 9$$
$$= 2 \times 3 \times 3 = 2 \times 3^2$$

3^2 is a shorthand for writing 3×3.

The prime factors of 18 are 2 and 3.

All numbers that have 2 as a factor are **multiples** of 2. ← *Numbers in the 2 times table*

All numbers that have 3 as a factor are multiples of 3. ← *Numbers in the 3 times table*

Squares, square roots and cubes

The square of 4 is $4 \times 4 = 4^2 = 16$.

16 is a **square** number.

The cube of 4 is $4 \times 4 \times 4 = 4^3 = 64$.

64 is a **cube** number.

If you want to find out what number when multiplied by itself is 36, you need to find the **square root** of 36. This is written as $\sqrt{36} = 6$.
In this case the answer is easy, but you might need to use your calculator to find the square root of a more difficult number.

Using a calculator ► page 10

1 Look at the numbers

6, 8, 10, 11, 12, 34, 49.

From this list write down

(a) the multiples of 4, (b) the square root of 64,
(c) the prime number.

MEG (SMP)

2 Write down the value of each of these.

(a) 5^2 (b) 2^3 (c) $\sqrt{9}$

3 3 4 5 6 8 9

From the list of numbers above, write down the numbers that are

(a) odd numbers, (b) square numbers, (c) multiples of 3.

MEG (SMP)

4 Put these numbers in order, smallest first:

3^2, 8^2, 3^3, 4^2, 5^3, 2^2.

5 Look at this list of numbers.

2 3 5 8 10 12 14 15

(a) Which of these numbers are (i) multiples of 5, (ii) prime numbers?

(b) Which numbers have 3 as a factor?

6 (a) What is the smallest number which is a multiple of 2 and 3?

(b) What four numbers are factors of 10?

7 (a) Write down all the prime numbers less than 10.

(b) Write down all the square numbers less than 20.

(c) Write down all the cube numbers between 5 and 10. ULEAC

8 This is a 'factor tree' for 24.

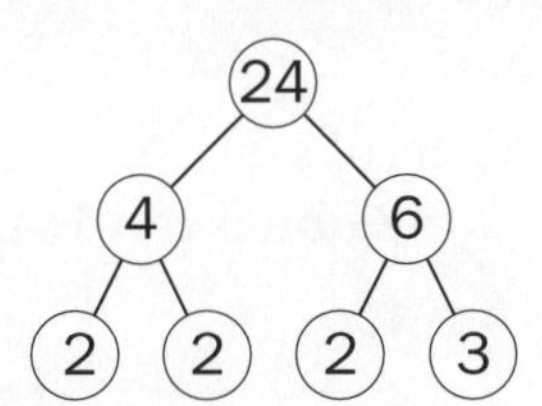

So $24 = 2 \times 2 \times 2 \times 3$
$= 2^3 \times 3$

Draw a factor tree to find the prime factors of 60.

9 3 4 9 13 15 23 25 28 64

Copy and complete these statements by choosing numbers from the list above to make these statements true.

(a) ...., ... and .... are prime numbers.

(b) .... and .... are factors of 200.

(c) .... is a multiple of 7.

(d) .... is a cube number.

(e) $5 \div$ .... $= \frac{1}{3}$. NICCEA

10 A pattern of counting numbers is shown.

14, 15, 16, 17, 18, 19, 20, ...

(a) (i) Which of these numbers is a square number?
(ii) Which of these numbers is a multiple of nine?

The pattern is continued.

(b) (i) What is the next square number?
(ii) What is the next number that is a multiple of nine? SEG

Answers and hints ► page 107

Ratio

Sally is making up orange drink in a café.
She mixes squash with water in the **ratio 1 to 6**.

> 1 to 6 is often written as **1 : 6**.

This means that for every 1 litre of squash she uses 6 litres of water.

Example

Tom is mixing up top dressing for a lawn.
For every 2 buckets of loam he adds 5 buckets of sand.
How many buckets of sand does he need to mix with 6 buckets of loam?

Here are three ways of working this out. You may have your own way.

(1) 2 buckets of loam need 5 buckets of sand, so
4 buckets of loam need 10 buckets of sand and
6 buckets of loam need 15 buckets of sand.

(2) 6 is 3 times as much as 2.
So Tom needs 3 times as much sand,
which is $5 \times 3 = 15$ buckets of sand.

(3) 2 buckets of loam mix with 5 buckets of sand, so
1 bucket of loam mixes with $5 \div 2 = 2\frac{1}{2}$ buckets of sand and
6 buckets of loam mix with $2\frac{1}{2} \times 6 = 15$

1 To make 12 toffee apples you need:

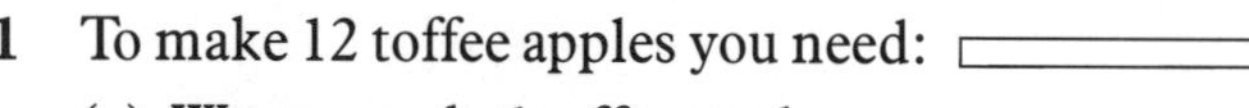

12 small eating apples
450g of sugar
50g of butter
1 tablespoon of golden syrup
2 teaspoons of vinegar
150ml of water

(a) Wayne made 6 toffee apples.
How much butter did he use?

(b) Patricia used 900 g of sugar.
How many toffee apples did she make?

MEG (SMP)

2 The fuel for a strimmer is a mixture of petrol and oil.
This has to be mixed in the ratio 25 to 1.
How much oil is needed to mix with 5 litres of petrol?

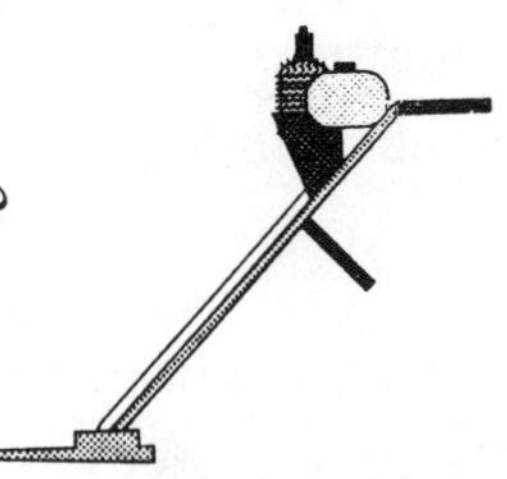

MEG (SMP)

3 A recipe for 24 mince pies uses 25 ounces of ready-made pastry and 20 ounces of mincemeat.

(a) Imran wants to make 12 mince pies.
How much pastry does he need?

(b) Stella uses 30 ounces of mincemeat.
How many mince pies does she make?

(c) Write down the ratio of mincemeat to pastry used in the recipe.

Sharing in a given ratio

A school raises £1800 with a school raffle.
It shares the money between itself and a charity in the ratio 3 to 1.
How much does each get?

Split the money into 4 equal parts. ⟹ $\frac{£1800}{4} = £450$

The charity gets 1 part, which is £450.

The school gets 3 parts, which is $£450 \times 3 = £1350$.

Check that the shares add up to £1800: £450 + £1350 = £1800

4 Jane and Alison win £600.
They divide the money in the ratio 3 : 2.

(a) How much money does Jane receive?

(b) How much money does Alison receive?

WJEC

5 Sophie takes part in a sponsored walk each year.
The money she raises is divided between two local charities, A and B, in the ratio 5 : 3.

(a) In 1996 she raised a total of £48.
How much did she give to charity A?

(b) In 1997 she gave £21 to charity B.
How much did she raise altogether?

MEG

6 Citrola drink is made by mixing grapefruit juice and pineapple juice in the ratio 3 : 2.

(a) At a factory the drink is mixed in large drums.
40 litres of pineapple juice are used.
How much grapefruit juice should be added?

(b) A bottle of Citrola holds 750 ml.
How much of this is pineapple juice?

MEG (SMP)

7 When John and Alec wrote a book they agreed to share any money they made from the sale of the book in the ratio 11 : 9.

(a) In 1993 they made £1520 from the sale of the book.
How much was John's share?

(b) In 1994 Alec's share was £450.
How much was John's share?

MEG (SMP)

More for your money

Which size toothpaste is the better value for money?

Small size

45 ml
85p

Medium size

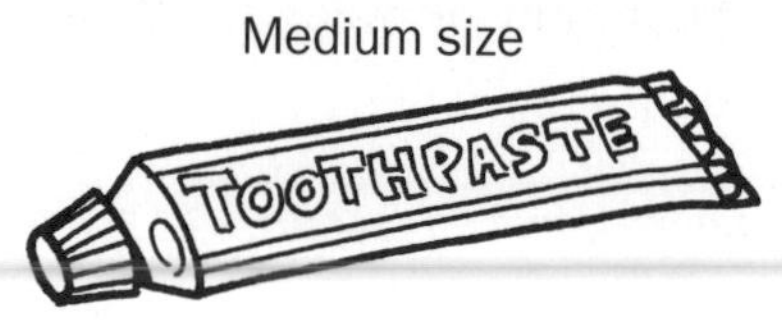

75 ml
135p

Here are two ways to compare value for money.

(1) Work out the cost per millilitre of each tube.

Small size

45 ml costs 85p.

1 ml costs $\frac{85}{45}$p = 1·89p (to 2 d.p.).

Medium size

75 ml costs 135p.

1 ml costs $\frac{135}{75}$p = 1·80p.

The medium tube is the better value for money: you pay **less pence per ml**.

(2) Work out how many millilitres you get for 1p for each tube.

85p buys 45 ml, so

1p buys $\frac{45}{85}$ml = 0·53 ml (to 2 d.p.).

135p buys 75 ml, so

1p buys $\frac{75}{135}$ml = 0·56 ml (to 2 d.p.).

The medium tube is the better value for money: you get **more ml per penny**.

8 A 200 ml can of SHINE car polish costs £3·30.
A 250 ml bottle of SHINE car polish costs £3·99.

(a) How many cans are needed for one litre of polish?

(b) Work out the cost of a litre of polish bought
(i) in cans, (ii) in bottles.

(c) Which is the better value for money?

MEG/ULEAC (SMP)

9 (a) Work out the cost of 100 g of *Bran Wheats* for each of these packets.

(b) Which packet gives the better 'value for money'?

(c) Give one possible reason why people might buy the other packet, even though it gives less 'value for money'.

MEG (SMP)

Answers and hints ► page 108

Mixed number

1 Work out which is the better deal for an apprentice who is currently paid £3·00 per hour.

40p an hour extra ***12% rise***

2 The Belmont hotel has 85 bedrooms.
A record was kept of the number of rooms occupied each night for a week.

Mon	Tue	Wed	Thur	Fri	Sat	Sun
55	60	69	74	60	55	40

What percentage of the rooms were occupied on Wednesday night?
Write your answer correct to 1 decimal place. MEG (SMP)

3 Mary earns £4·20 per hour.

(a) Calculate how much she earns in a 38-hour week.

(b) For each hour over 38 hours she works, Mary is paid overtime.
The overtime rate is 'time and a quarter'.
How much is she paid for each hour of overtime?

(c) Mary is given a rise of 5%.
Calculate how much she earns in a 38-hour week after the rise.

MEG (SMP)

4 Mr and Mrs Evans have just received their electricity bill for the April to June quarter.
The details of the bills are as follows:

Previous meter reading	88 156
Present meter reading	89 343
Cost per unit is 8 pence	
Fixed standing charge per quarter	£11·05

(a) Find the **total** cost of the electricity for the April to June quarter, giving all the details of your calculations.

(b) VAT of 5% is charged on electricity bills.
How much is Mr and Mrs Evans' electricity bill including VAT? WJEC

5 Sandra has five cards with numbers on them as shown.

7 **8** **1** **6** **4**

(a) Write down the smallest **three-digit** number Sandra can make using these cards.

(b) (i) Write down the largest **four-digit** number Sandra can make using these cards.

(ii) Write your answer to (i) correct to the nearest thousand.

6 **You must not use a calculator for this question.**

Beacon School is organising a trip to a theme park.
307 pupils have each paid £19 for the trip.

(a) Gina wants to know roughly how much the pupils have paid altogether.
Write down the calculation Gina could do in her head, and work it out.

(b) Work out the exact amount that has been paid altogether for the trip.

(c) The school is using coaches for the trip.
Each coach will seat 48 pupils.
How many coaches will be needed?

MEG (SMP)

7 (a) What fraction of the diagram is shaded?

(b) What percentage of the diagram is shaded?

(c) Stuart adds some shading so that the ratio of unshaded squares to shaded squares is 3 : 2.
How many more squares must he shade?

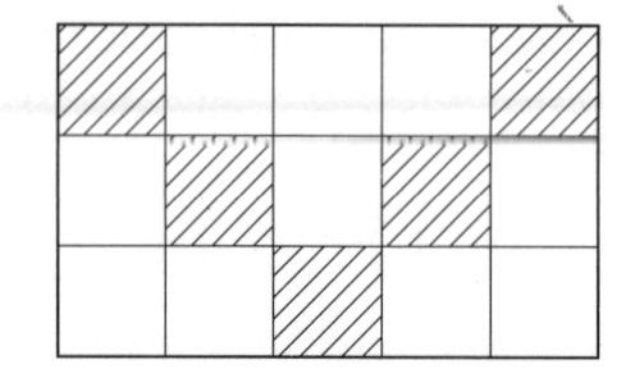

8

Households with television licences

	1966	1971	1976	1981	1986
Black and white	53%	62%	36%	18%	12%
Colour	0%	10%	49%	69%	81%

To watch TV, you must have either a black and white licence or a colour licence.

(a) What percentage of households had a TV licence in 1976?

(b) Comment on how the percentage of licences for colour television compares with the percentage of licences for black and white television over the years 1966 to 1986.

(c) In 1991, 97% of households had a TV licence.
What fraction of households did not have a TV licence?

MEG

9 (a) Which of the numbers greater than 1 but less than 6 are prime numbers?

(b) (i) What is the value of 7 in the number 1670?

(ii) The number 1670 is multiplied by 10.
What is the value of 7 in the answer?

(iii) The number 1670 is divided by 10.
What is the value of 7 in the answer?

(c) What number less than 10 is a multiple of 2 and can be divided exactly by 3?

10 The size of a crowd at a rugby match is given as 12 600 to the nearest hundred.

(a) What is the largest number the crowd could be?

(b) What is the smallest number the crowd could be?

(c) The ratio of men to women is about 2 : 1.
Approximately how many of the spectators are women?

11 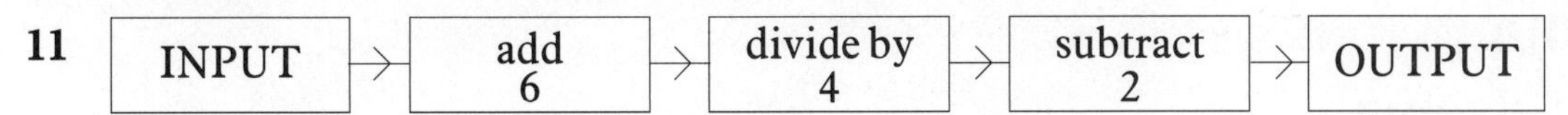

Use the above instructions to copy and complete this table.

INPUT	OUTPUT
22	5
3	
$^-2$	
	10

WJEC

12 Mrs Keetch's washing machine has gone wrong.
She calls out a plumber.
The charge for replacement parts is £29·75.
Copy and complete this bill.

O.K. PLUMBERS	
Fixed call-out fee	£21·50
$1\frac{3}{4}$ hours at £16 per hour	
Parts	£29·75
Total before VAT	
VAT at $17\frac{1}{2}$%	
Total due	

13 David is a scout leader.
He is calculating the cost of a day trip to a theme park for 182 scouts.

(a) They will travel by coach.
Each coach can carry 48 people.
How many coaches will he need to hire?

The cost of entry to the theme park is £13 per person.

(b) (i) David uses a calculator to work out the total cost of entry for 182 scouts.
Write down, in order, the keys David should press.

(ii) David's calculator does not work. Calculate, by writing down in full, how he works out the total cost **without** using a calculator.

For scout groups the theme park gives a 15% discount on the £13 cost of entry.

(c) What is the cost of entry for each scout?

SEG

14 Paul bought a new car priced £7500.
He paid $\frac{2}{5}$ of this price when he bought it.
He paid the rest by equal monthly payments for the next 12 months.

(a) How much did he pay when he bought the car?

(b) What was his monthly payment for the next 12 months?

MEG (SMP)

Answers and hints ► page 109

ALGEBRA

Writing algebra

Using shorthand

Anne buys 3 oranges and Rob buys 7 oranges.
The oranges cost 25p each.
The total cost of the oranges in pence is

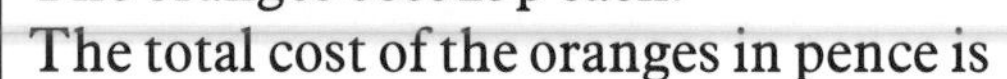

$$(3 \times 25) + (7 \times 25) = 10 \times 25$$
$$= 250$$

(3 lots of 25) + (7 lots of 25) = 10 lots of 25

If the oranges had cost a pence each then we could write

$$(3 \times a) + (7 \times a) = 10 \times a$$
$$= 10a$$

(3 lots of a) + (7 lots of a) = 10 lots of a

Instead of writing $10 \times a$ we usually write $10a$.

Suppose Anne also buys 7 grapefruit for b pence each.
Then the total cost of Anne's shopping in pence is $3a + 7b$.

Notice that you cannot add 3a and 7b together because they are not ***like*** *terms.*

Here are some more shorthand expressions that you need to understand.

ab means $a \times b$ $\frac{a}{2}$ means $a \div 2$ $\frac{a}{b}$ means $a \div b$

$a^2 = a \times a$ $a^3 = a \times a \times a$ $ab + c$ means $(a \times b) + c$

Using formulas and expressions ► page 35

1 Simplify these algebraic expressions, where possible.

(a) $a + 2a$ (b) $b + 4b - 3b$ (c) $d + 5d - 4$ (d) $2 + a + 1$
(e) $3m + 2$ (f) $5x + 7 - 3x$ (g) $c + 7d + 3c$ (h) $2q + 4r - q$
(i) $2 \times 3 \times a$ (j) $5 \times 6c$ (k) $4c \times c$ (l) $2 \times c \times d$
(m) $a^2 \times a$ (n) $3a \times 2b$ (o) $x \times y \times 5$ (p) $2x \times x^2$

2 Write down expressions for each of the lengths marked ?.

(a)

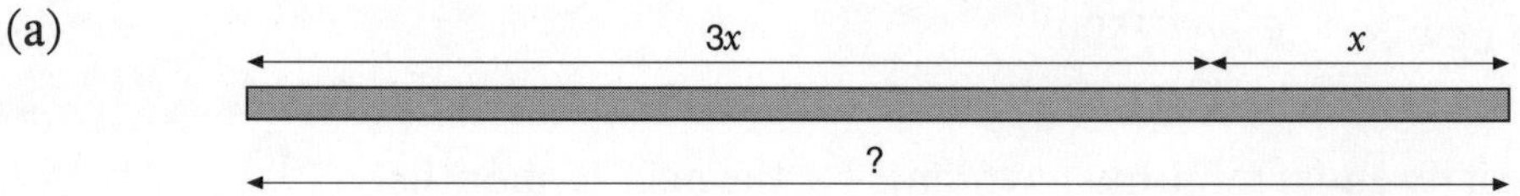

(b)

z
w ?

MEG/ULEAC (SMP)

3 Long straws are x cm long.

Short straws are y cm long.

Write down as simply as possible, using x and y, the lengths of these lines of straws.

(a)

(b)

MEG/ULEAC (SMP)

4 (a) Zahida has c one pound coins. Asma has d one pound coins.
How many pound coins have they got altogether?

(b) Bill has m pounds. He spends n pounds.
How much money has he left?

MEG (SMP)

5 Ben works for 7 hours and is paid £a per hour.
How much is he paid altogether?

6 Write down an expression for the area of each of these.

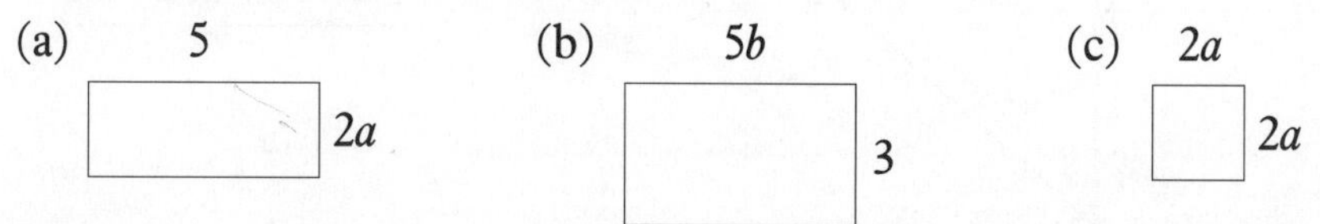

Removing brackets

$3(x + 5)$ is the same as $(3 \times x) + (3 \times 5)$, that is $3x + 15$.

*This 3 outside the brackets means 'multiply the x **and** the 5 by 3'.*

Example Simplify $4(2x - 1) - 3x$.

Multiply both items inside the brackets by 4. $(4 \times 2x) - (4 \times 1) - 3x = 8x - 4 - 3x$

Collect like terms together. $= 5x - 4$

7 Remove the brackets from these expressions and simplify them where possible.

(a) $4(c + 6)$ (b) $3(d - 4)$ (c) $2(2a + 1) + 3$ (d) $5(6 - x) - 10$

(e) $a + 3(a - 4)$ (f) $5(x + 1) - 3$ (g) $4(3 + 2x) - x$ (h) $2(3a + 4b) + 1$

8 (a) Marcus is paid £x per hour for work done during Monday to Friday.
How much was he paid for 30 hours of work during Monday to Friday?

(b) On Saturday he gets paid at twice the Monday to Friday rate.
How much was he paid for working a total of 8 hours on Saturday?

(c) How much was Marcus paid altogether? Give your answer in its simplest form.

(d) Christopher is paid £2 per hour more than Marcus.
How much was Christopher paid for 6 hours worked on a Monday?

WJEC

Answers and hints ► page 110

Coordinates and graphs

Coordinates

Check that you agree that these are the **coordinates** of the corners of the triangle ABC:
A is the point (4, 2), B is the point ($^-3$, 2) and C is the point ($^-3$, $^-4$).

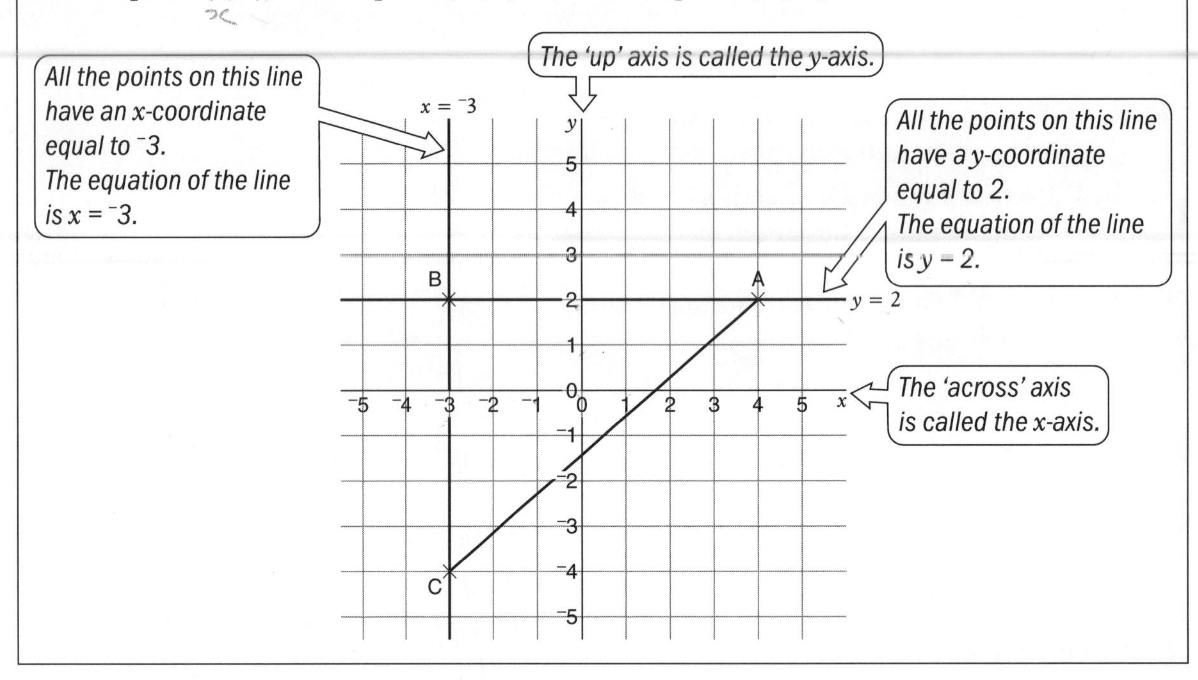

1 Copy the axes on the right on to squared paper and mark the point A.

(a) What are the coordinates of A?

(b) Plot the point (2, 1) and label it B.

(c) Plot the point (1, $^-1$) and label it C.

(d) (i) Plot the point D to make ABCD a rectangle.
(ii) What are the coordinates of D?

MEG/ULEAC (SMP)

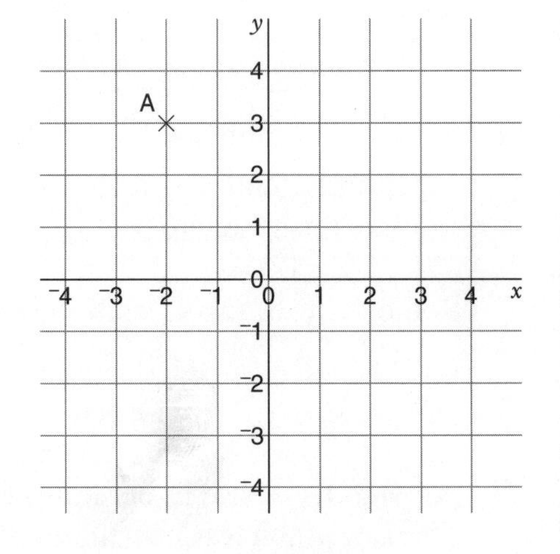

2 Make another copy of the axes in question 1.

(a) Mark and label the points P (3, 4) and Q ($^-4$, $^-3$).

(b) Joint the points P and Q with a straight line.
(i) Write down the coordinates of the point where the line cuts the x-axis.
(ii) What are the coordinates of the mid-point of the line PQ?

(c) Draw and label the line with equation $y = {}^-3$.

Drawing graphs

Example

(a) Complete the table of values and draw the graph of $y = x^2 + 1$.

x	0	1	2	3	4	5	6
y			5		17		37

(b) Use your graph to find the value of x when $y = 30$.

Substituting ► page 35

Solution

(a) Use the equation to calculate the unknown values of y.

When $x = 0$, $y = 0^2 + 1 = 1$
When $x = 1$, $y = 1^2 + 1 = 2$
When $x = 3$, $y = 3^2 + 1 = 10$
When $x = 5$, $y = 5^2 + 1 = 26$

This is the completed table.

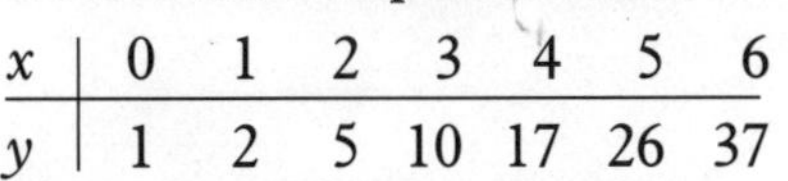

x	0	1	2	3	4	5	6
y	1	2	5	10	17	26	37

Then plot the points from the table and draw a smooth curve through them.

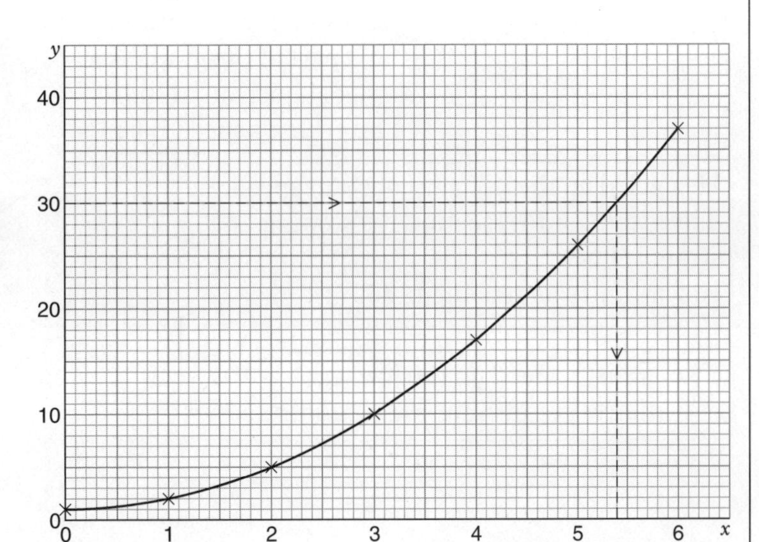

(b) The arrowed lines show how to find the value of x when $y = 30$.
Check you agree that $x = 5{\cdot}4$ when $y = 30$.

Time and conversion graphs ► page 32

Line graphs ► page 71

3 (a) Copy and complete this table of values for $y = 4x + 3$.

x	0	1	2	3	4	5
y		7		15		

(b) Draw the graph of $y = 4x + 3$. (Start by copying the axes above on 2 mm graph paper.)

(c) On the same diagram draw the line with equation $x = 2$.

(d) What are the coordinates of the point where the two lines cross?

4 The graph of $y = 2x^2$ can be drawn by plotting the points in this table and joining them with a smooth curve.

x	0	1	2	3	3·5	4
y	0	2	8	18	24·5	32

(a) Draw the graph of $y = 2x^2$. (Start by copying the axes above on 2 mm graph paper.)

(b) On the same axes draw the straight line $y = 5x$.

(c) What are the coordinates of the points where the line cuts the curve?

Questions 5 and 6 are on worksheets F1 and F2.

Answers and hints ► page 111

Time and conversion graphs

Graphs that show how a quantity varies over a period of time can be used to 'tell a story' about what is happening.

A man cycles up a hill, rests at the top and then cycles down the other side.

The graph shows the distance he cycled during these three stages of his journey.

He rested for 3 minutes. This part of the graph is a straight horizontal line. The cyclist's speed is 0 m/s!

This line slopes steeply. It shows that the cyclist went faster down the hill than up.

He cycled up the hill at a constant speed. The graph for this section is a sloping straight line. In 5 minutes he travelled 1200 m.

The cyclist's speed up the hill was $\frac{1200}{5}$ metres per minute = 240 m/min.

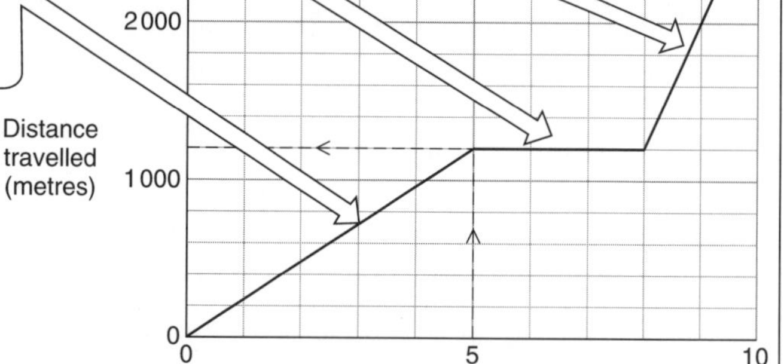

Rates of change ► page 58

Line graphs ► page 71

1 This distance–time graph shows a train journey from Birmingham one afternoon.

(a) What happened at 2:15 p.m.?

(b) Between which two times did the train travel fastest?

(c) (i) How long did the whole journey take?

(ii) How many miles did the train travel altogether?

(iii) What is the average speed for the whole journey in miles per hour?

MEG (SMP)

Average speed ► page 58

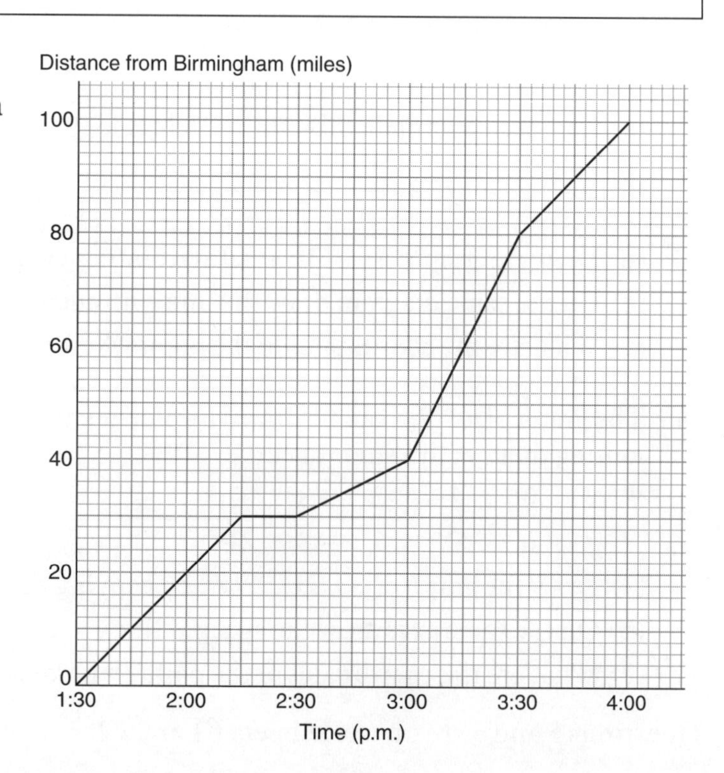

2 This sketch graph shows what happened to the volume of water in a bath.
At the start point, A, the plug was in the bath and the hot tap was turned on.

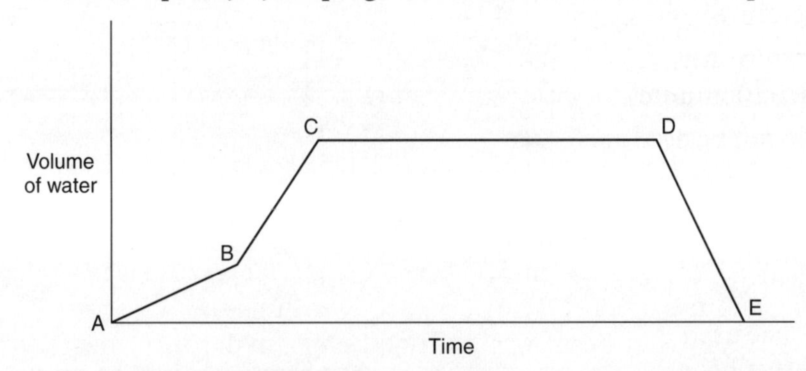

(a) Which point on the graph shows when the plug was pulled out?

(b) What do you think happened at B?

(c) What happened at C?

MEG (SMP)

3 This graph helps to convert between pounds (£) and dollars ($).

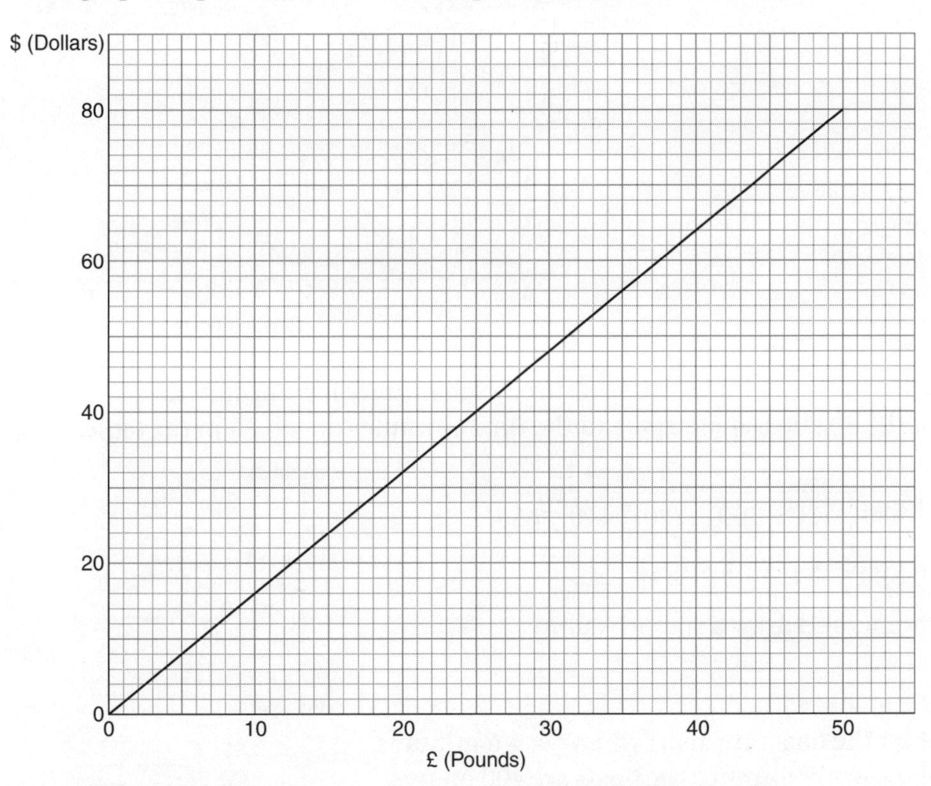

(a) Before travelling to New York, Ken changes £40 into dollars.
How many dollars does he get?

(b) A meal costs $24.
How much is this in pounds?

MEG (SMP)

4 The graph shows the temperature of a photographic solution which cools gradually from a temperature of 60°C.

(a) What is the temperature after
  (i) $7\frac{1}{2}$ minutes (ii) 10 minutes?

(b) The solution should not be used at a temperature above 25°C.

  How long does the solution stand before a temperature of 25°C is reached?

(c) The solution cannot be used when its temperature is below 18°C.

  When does this happen?

(d) For how long is the solution usable?

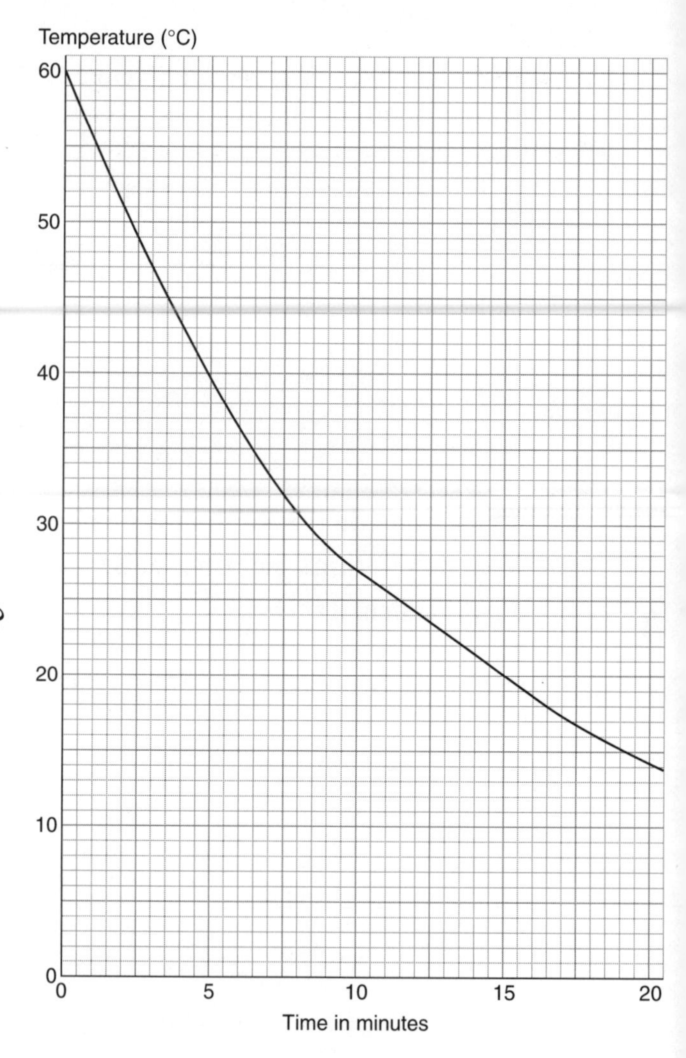

5 The distance between two boats was measured each minute for 5 minutes.
The values of the distance, d metres, and the time t minutes are given in the table.

t (minutes)	0	1	2	3	4	5
d (metres)	880	530	300	280	440	650

Draw axes on graph paper like these.

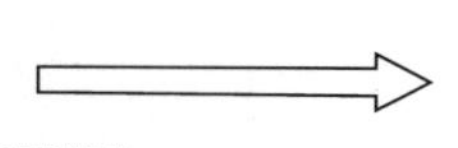

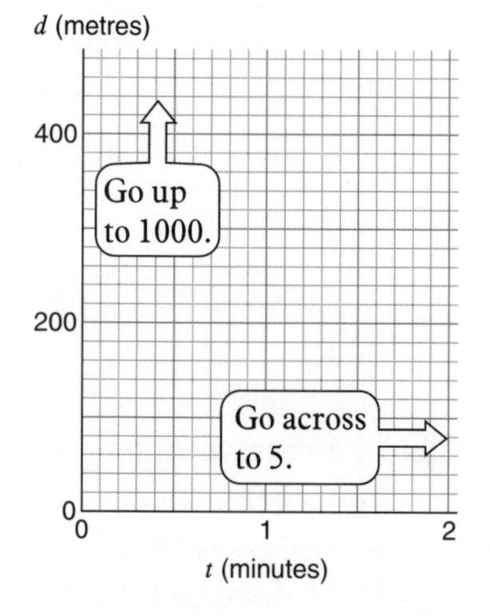

(a) Plot the points and draw a smooth curve through them.

(b) From your graph, estimate
  (i) how far the boats are apart when $t = 4{\cdot}6$ minutes,
  (ii) for how many minutes the boats are 400 metres or less apart,
  (iii) the values of t and d when the boats are closest to one another.

NICCEA

Question 6 is on worksheet F3.

Answers and hints ► page 112

Using formulas and expressions

Substituting numbers into an expression

Examples

In these examples we will use the values $a = 2, b = 5$ and $c = 3$ to work out the values of **expressions** such as $2a + 3b$.

$2a + 3b = (2 \times 2) + (3 \times 5) = 4 + 15 = 19$ ⇐ Work out $(2 \times a)$ and $(3 \times b)$ and add them.

$ab - c = (2 \times 5) - 3 = 10 - 3 = 7$ ⇐ Work out $(a \times b)$ and subtract c.

$\frac{b + c}{a}$ means $(b + c) \div a = (5 + 3) \div 2 = 8 \div 2 = 4$ ⇐ Work out $(b + c)$ and then divide by a.

$c^2 + 2 = (3 \times 3) + 2 = 9 + 2 = 11$ ⇐ Work out $(c \times c)$ and then add 2.

Writing algebra ► page 28

Writing formulas

This shape is made from rods of length l and rods of length s.

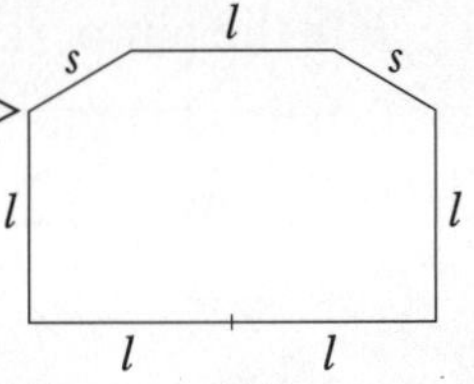

The perimeter, P, (the distance round the outside) is $l + s + l + l + l + l + s$.
This expression can be written more simply as $5l + 2s$.

The **formula** for the perimeter is

$$P = 5l + 2s.$$

Suppose that $l = 12$ and $s = 8$.
We can **substitute** these numbers into the formula to find the value of P.

$P = (5 \times 12) + (2 \times 8)$ ⇐ When you replace letters by numbers, put brackets in.
$= 60 + 16$
$= 76$

Forming and solving your own equations ► page 39

1 If $a = 5, b = 2$ and $c = 3$, work out the value of these expressions.

(a) $2b$ (b) $a + b$ (c) $a - c$ (d) $3a + 5$ (e) b^2
(f) $4c - 6$ (g) $4b + 3c$ (h) $7c - 4a$ (i) $2a^2$ (j) $c^2 - 1$

2 Here is a number machine chain.

Input → ×2 → +5 → Output

(a) (i) The input is 3. What is the output?
(ii) What is the output for an input of a?

(b) Copy this number machine chain.

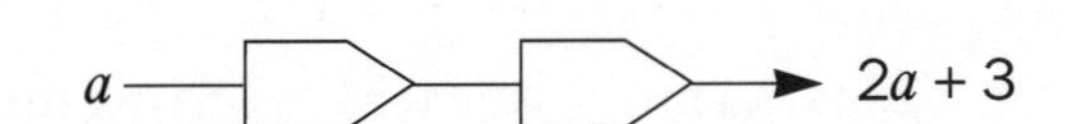

Complete the number machine chain so that it will give an output of $2a + 3$ for an input of a.

3 If $a = 3, b = 4$ and $c = 6$, work out the values of these expressions.

(a) $\frac{c}{a}$ (b) $\frac{4a}{c}$ (c) $ab + c$ (d) $\frac{ac}{2}$ (e) $\frac{3b}{a}$

(f) a^3 (g) $\frac{2b - a}{5}$ (h) $a^2 + b^2$ (i) $b + \frac{c}{a}$ (j) $a(b + 1)$

4 A table has a circular glass top. The radius of the glass top is 40 cm.
The formula $A = 3 \times r \times r$ gives you the approximate area of a circle.

r is the radius in centimetres.
A is the area in square centimetres.

Use this formula to work out the approximate area of the glass top. MEG (SMP)

5 A rough formula for conversion of temperatures is

$F = 2C + 30.$

C is the temperature in degrees Celsius.
F is the temperature in degrees Fahrenheit.

(a) Find F when $C = 18$.

(b) Find F when $C = 0$. MEG (SMP)

6

Phil is cooking a chicken for Sunday lunch.
He looks in a cookery book for some instructions.

It says

'Cook for 20 minutes per pound
plus an extra 30 minutes.'

Write this instruction as a formula connecting t and w.

t is the cooking time in minutes,
w is the weight of the chicken in pounds. MEG (SMP)

7 Divers use this formula to work out how long they can stay under water.

$$\text{time} = \frac{12 \times \text{volume of air in tank}}{\text{depth}}$$

time is in minutes,
volume of air in tank is in cubic feet,
depth is in feet.

Chris has 30 cubic feet of air in her tank.
She is at a depth of 20 feet.
Use the formula to work out how long she can stay under water.

MEG (SMP)

8 This is a rough rule for converting miles to kilometres.

$k = m \times 1{\cdot}6,$

where k is the number of kilometres and m is the number of miles.

Use this rule to convert 70 miles to kilometres.

9 The formula $t = \frac{d}{s}$ is used to work out the time for a journey.

t is the time in hours,
d is the distance in miles,
s is the average speed in m.p.h.

Daniel expects to do a journey of 330 miles at an average speed of 55 m.p.h.
Use the formula to work out an estimate of his journey time.

10

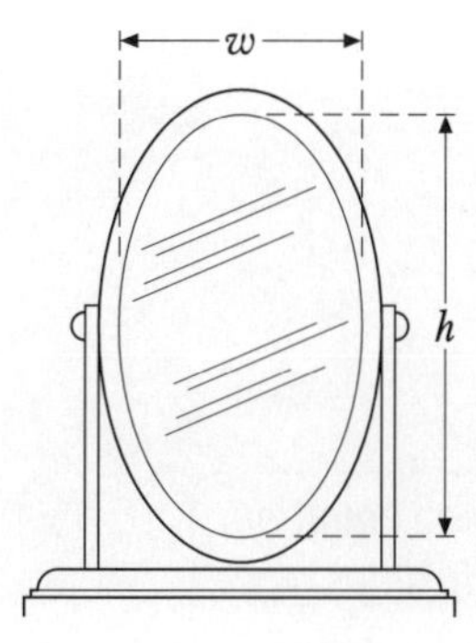

There is an approximate formula for the area of an oval.
It says that the area, A, can be found by multiplying the height, h, by the width, w, and multiplying the result by 0·8.

(a) Write this formula for the area of an oval using the letters A, w and h.

(b) An oval mirror is 10 cm wide and 15 cm high.
Find its area.

MEG (SMP)

11 The number of lines, N, used to draw a Mystic Rose can be calculated using the formula

$$N = \frac{p(p-1)}{2}$$

where p is the number of points on the circle.
Use the formula to calculate the number of lines needed to draw a Mystic Rose with 15 points on the circle.

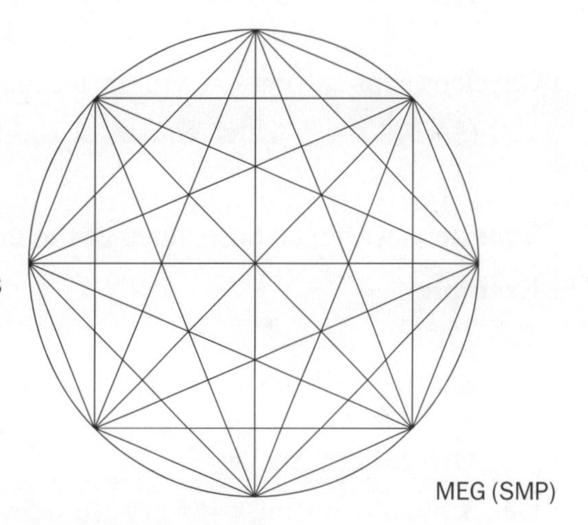

MEG (SMP)

12 The formula $V = Ah$ can be used to work out the volume of a cylinder.

V is the volume in cm^3,
A is the area of the cross-section in cm^2,
h is the height in cm.

$A = 24·4\,cm^2$
$h = 10·5\,cm$

Use the formula to work out the volume of this cylinder.

MEG (SMP)

Answers and hints ► page 114

Solving linear equations

Equations with x on one side of the equation only

When you **solve** the equation $3x - 2 = 7$ you are finding the value of x that will make $3x - 2$ equal to 7.

The value of x for which this is true is called the **solution** to the equation.

To solve (find the solution of) an equation we use the fact that an equation is still true if

- the same number is added to (or subtracted from) each side or
- each side is multiplied (or divided) by the same number.

Example

Solve $3x - 2 = 7$.

$$3x - 2 = 7$$

Add 2 to each side. $3x - 2 + 2 = 7 + 2$

$$3x = 9$$

Divide both sides by 3. $3x \div 3 = 9 \div 3$

$$x = 3$$

When you solve an equation, always keep both sides balanced.

Check by substituting $x = 3$ in the left-hand side of the original equation:

$(3 \times 3) - 2 = 7$, which equals the right-hand side.

Substituting numbers ► page 35

Equations with x on both sides of the equation

Example

Solve $5x - 7 = 2x + 5$.

Subtract $2x$ from both sides. $3x - 7 = 5$

Add 7 to both sides. $3x = 12$

Divide both sides by 3. $x = 4$

Check that you understand this working.

Check by substituting $x = 4$ in both sides of the equation.

$5x - 7 = 20 - 7 = 13$ and

$2x + 5 = 8 + 5 = 13$

So $x = 4$ is the correct value.

1 Solve each of these equations.

(a) $4x = 32$ (b) $x + 7 = 15$ (c) $x - 8 = 20$

(d) $2x + 1 = 9$ (e) $3x - 4 = 11$ (f) $7x - 14 = 0$

(g) $22 = 5x + 2$ (h) $2(x - 3) = 1$ (i) $5(2x + 1) = 30$

2 Solve each of these equations.

(a) $2x + 1 = x + 9$ (b) $3x - 7 = x + 5$ (c) $5x - 9 = 2x + 6$

(d) $2 + 5x = 10 - 3x$ (e) $10 + 2x = x + 14$ (f) $2(x + 1) = x + 5$

Forming and solving your own equations

Saria is thinking of a number.
If she multiplies it by 2 and adds 5 she gets 17.
What number is Saria thinking of?

Let the number be n.

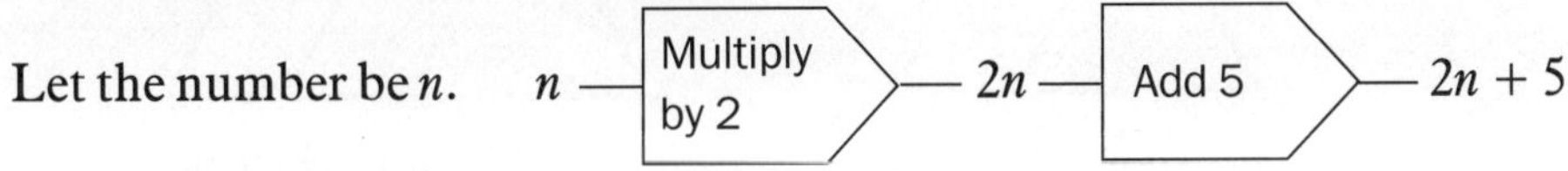

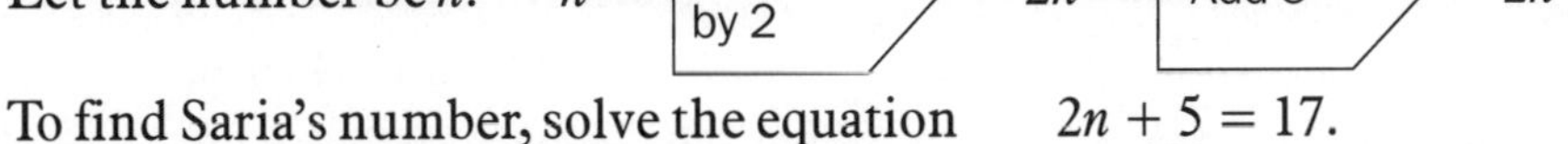

To find Saria's number, solve the equation $2n + 5 = 17$.

Subtract 5 from both sides. $2n = 17 - 5 = 12$

Divide both sides of the equation by 2. $n = 12 \div 2 = 6$

Check by substituting back in the **original explanation**: $(6 \times 2) + 5 = 17$
Do not use your equation in case it is wrong.

3 I think of a number, n, multiply it by 4, subtract 3 and get an answer of 13.
Write this down as an equation and solve it to find the number I first thought of.

4 The sum of the sides of this triangle is 22 cm.

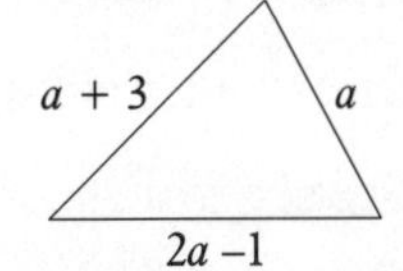

(a) Write an equation in terms of a.
Solve it to find the value of a.

(b) Write down the lengths of the other two sides.

5 When a number is multiplied by 3 and 5 is added to the result, the answer is 23.
What is the number?

6 The cost, C pence, of advertising in the local newspaper is worked out using the formula

$$C = 20n + 30,$$

where n is the number of words in the advertisement.

(a) Annalise puts in an advertisement of 15 words. Work out the cost.

(b) The cost of Debbie's advertisement is 250 pence.
 (i) Use the formula to write down an equation in n.
 (ii) Solve the equation to find the number of words in Debbie's advertisement. MEG

7 The angles of a triangle are x, $2x$ and $3x$.

(a) Write down an equation in x and solve it to find the value of x.

(b) Use this to write down the three angles of the triangle.

Angle sum of a triangle ► page 50

8 There were some cars in a car park.
During the day 55 more cars parked but none left.
There were then 6 times as many cars as before.
How many cars were in the car park at first?

Answers and hints ► page 115

Sequences and terms

Look at these patterns of rods.

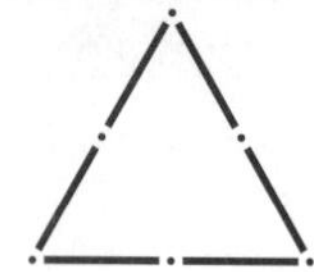

Pattern 1 has 6 rods. Pattern 2 has 10 rods. Pattern 3 has 14 rods.

We can write the numbers of rods as a **sequence**.

6, 10, 14, 18, 22, ...

Check you agree that the next two terms are 18 and 22.

Suppose you wanted to find out how many rods there were in pattern 10.
You could count on in fours or you might notice that each term is

(4 times its pattern number) + 2.

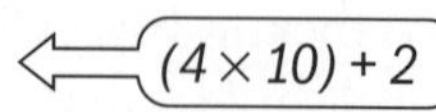

So pattern 10 has 42 rods. *(4 × 10) + 2*

Similarly, pattern 12 has 50 rods *(4 × 12) + 2*

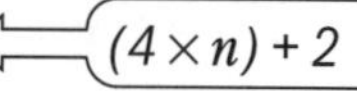

and pattern n has $4n + 2$ rods. *(4 × n) + 2*

It is often easier to see from a table of values how a sequence is built up:

Pattern	1	2	3	4	5	...	n
Number of rods	6	10	14	18	22		$4n + 2$

1 The diagram shows some patterns made with matches.

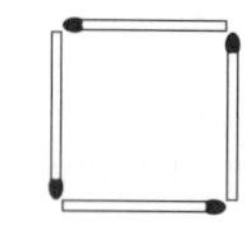

Pattern 1

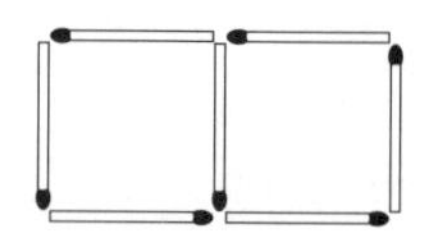

Pattern 2

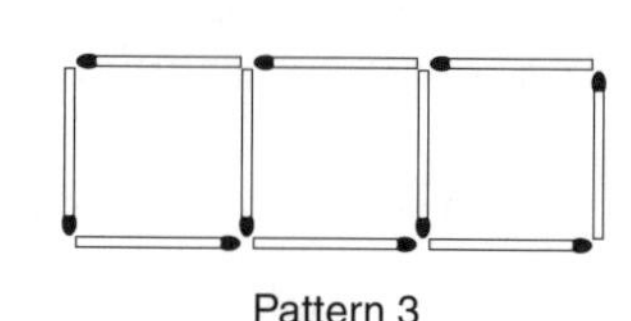

Pattern 3

(a) Sketch the 4th and 5th patterns in the sequence.

(b) Copy and complete this table.

Pattern	1	2	3	4	5
Number of matches	4	7			

(c) Which pattern can be made with exactly 25 matches?

(d) Explain how you could work out the number of matches needed for pattern 16 without doing any drawing.

MEG (SMP)

2 The first three terms of a sequence are

1, 4, 9, ...

The rule for finding the next number in the sequence is '**Add on the next odd number**'.

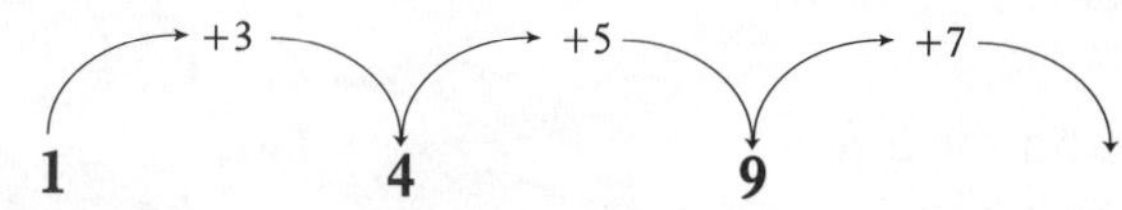

(a) Write down the next three numbers in the sequence.

(b) What is the special name for these numbers?

MEG (SMP)

3 Simon is stacking toy bricks.

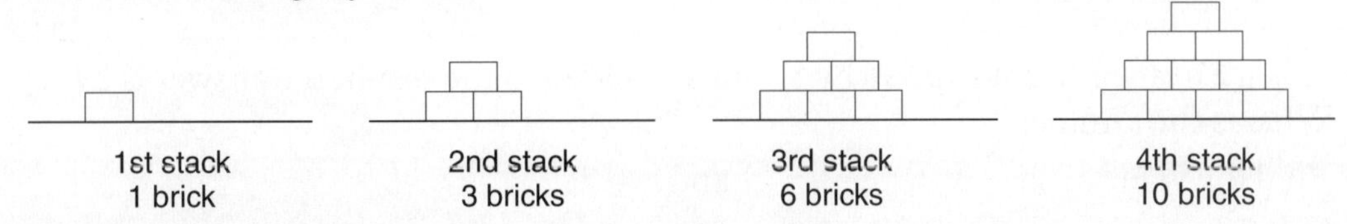

(a) How many bricks will there be in the 5th stack?

(b) Simon makes a stack with 28 bricks in it.
What is the stack number?

MEG (SMP)

4 (a) Write down the next two numbers in each of these sequences.

(i) 1, 6, 11, 16, … (ii) 1, 2, 4, 8, 16, …
(iii) 27, 24, 21, 18, …

(b) For each sequence write down your rule for finding the next number.

5 6, 12, 18, 24, …

(a) Write down the next two terms in the sequence above.

(b) Explain how you could write down the 10th term in the sequence without writing down all the terms.

(c) (i) What is the 20th term in the sequence 4, 8, 12, 16, … ?
(ii) What is the nth term in this sequence?

6 (a) Look at this sequence of numbers:

3, 8, 18, 38, …

The rule that has been used to get each number from the number before it is

Add 1 and then multiply by 2.

(i) Write down the next number in this sequence.
(ii) Using the same rule but a different starting number, the second number is 16.
What is the starting number?

(b) Look at this sequence of numbers:

5, 9, 13, 17, 21, …

(i) Write down, in words, the rule for getting each number from the one before it.
(ii) Write down a formula, in terms of n, for the nth number in the sequence.

MEG

7 (a) For each of these sequences, write in words the rule for getting each number from the one before it.

(i) 3, 6, 9, 12, … (i) 3, 5, 7, 9, … (iii) 4, 9, 14, 19, …

(b) What is the nth term for each sequence?

Answers and hints ► page 116

Mixed algebra

1 (a) When a number is multiplied by 5, and 4 is added to the result, the answer is 29. What is the number?

(b) Find the value of $2a + 3b$ when $a = {}^-1$ and $b = 4$.

(c) Simplify

(i) $2a + 3b - 4a$, (ii) $3x + 2(2x - 3y)$.

WJEC

2 *Bonus* coachline have a 50 seater coach for hire. They use the formula

$$C = 100 + 2m$$

to work out the charge (£C) to hire the coach to travel m miles in one day. Ecclesall sports club hired the coach to travel 160 miles in one day.

(a) Work out the cost of hiring the coach.

To go on the journey the club charged adults £10 and children £5. 35 adults and 10 children paid.

(b) How much did they pay altogether?

(c) (i) Say whether the club made a profit or a loss.

(ii) How much was this profit or loss?

MEG/ULEAC (SMP)

3 A rectangle has a length of a cm and a width of b cm. The perimeter of a rectangle is given by the formula $p = 2(a + b)$. Calculate the perimeter of a rectangle when $a = 4{\cdot}5$ and $b = 4{\cdot}2$.

SEG

4 The formula connecting the length of a car's skid and its speed before starting to skid is:

$$L = 0{\cdot}02\,S^2$$

where L is the length of the skid in metres,
S the speed of the car in miles per hour.

(a) Work out L when S is 100 miles per hour.

(b) In 1964 the jet-powered car *Spirit of America* went out of control on the salt flats in Utah, America.
It made a skid mark 9600 metres long!
Use trial and improvement to find to the nearest 1 mile per hour the speed the car was travelling at before it skidded.
You may find a table like this useful. (You will need to extend it.)

Speed (S miles per hour)	S^2	$0{\cdot}02S^2$	Trial too small	Trial too big
600	360000	7200	✓	
700	490000	9800		✓

MEG (SMP)

5 A sixth form young enterprise group is making money by selling T-shirts.

The charge, £p, for s T-shirts is given by the formula

$$p = 5s + 25.$$

(a) Copy this table and use the formula to complete it.

s	0	20	40	60	80
p					

(b) Draw a graph with axes like these.

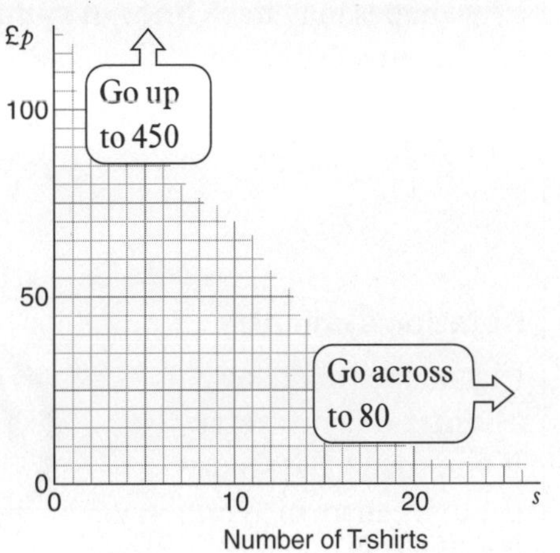

(c) (i) Use the graph to find the cost of buying 50 T-shirts.

(ii) If 50 T-shirts are bought, what is the price per T-shirt?

(d) Use the graph to find how many T-shirts you can buy for £350.

6 Here is a table, showing a pattern of addition sums.

(a) Copy and complete line 5 of the table.

(b) Write down line 6 of the table.

(c) What special name is given to the numbers

(i) 1, 3, 5, 7, ... ,

(ii) 1, 4, 9, 16, 25, ... ?

(d) Write down, in terms of n, the nth term of each of the sequences in part (c).

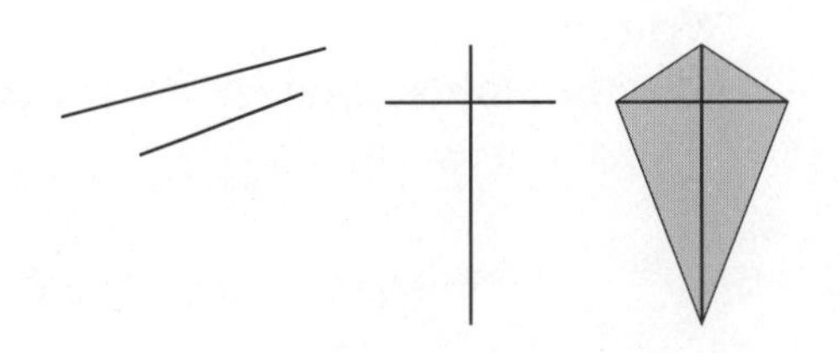

Line 1	1	= 1
Line 2	1 + 3	= 4
Line 3	1 + 3 + 5	= 9
Line 4	1 + 3 + 5 + 7	= 16
Line 5	1 + 3 + 5 + ... + ...	= 25
Line 6		=

MEG/ULEAC (SMP)

7 A kite frame is made from two sticks.
The sticks are fixed together at right-angles.
Material is then fixed to the sticks to make a kite.
The formula

$$A = 0{\cdot}5ef$$

gives the area of the kite.
A is the kite's area in square centimetres.
e and f are the lengths of the two sticks in centimetres.

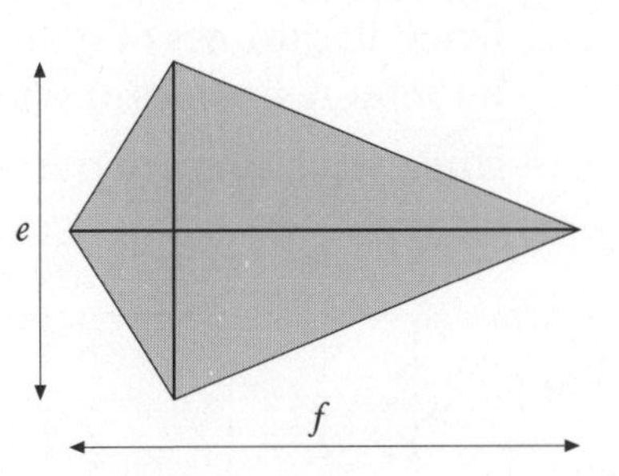

(a) What is the area of a kite with sticks 40 cm and 90 cm long?

(b) Mark wants to make a kite with an area of 3000 cm^2.
He has one stick which is 100 cm long.
What length should the other stick be?

MEG/ULEAC (SMP)

Answers and hints ► page 117

SHAPE, SPACE AND MEASURES

Understanding shape

Reflection symmetry

Flat shapes may have **lines of symmetry**.
If you put a mirror on a line of symmetry the shape will look the same.

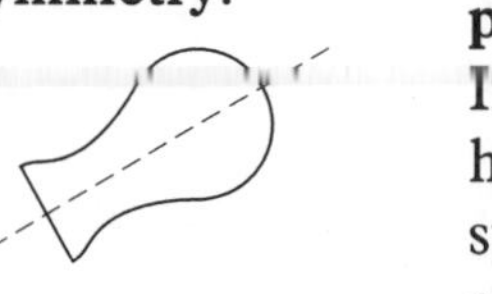

Solid shapes may have **planes of symmetry**.
If you chop the solid in half down a plane of symmetry and put half on a mirror, it will look the same.

Rotation symmetry

Flat shapes may have a **centre of symmetry**.
If you turn a flat shape about a centre of symmetry it will fit on to itself.
The number of different positions it will fit is called the **order of rotation**, or **order of rotation symmetry**.
This shape has rotation symmetry of order 2.

Solid shapes may have **an axis of symmetry**.
If you turn a solid shape about an **axis of symmetry**, it will fit onto itself.
The number of different positions it will fit is called the **order of rotation**.
This solid has order of rotation symmetry 4 about this axis.

Naming shapes

You should be able to recognise these shapes and name them:

isosceles triangle, scalene triangle, equilateral triangle, square, rectangle, trapezium, parallelogram, rhombus, kite, pentagon, hexagon.

1 Write down the name of each of these shapes.
Give the most accurate name you can.

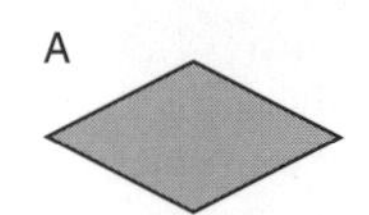

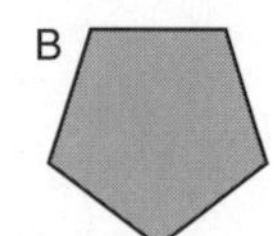

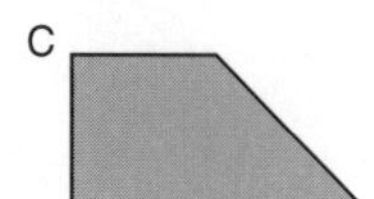

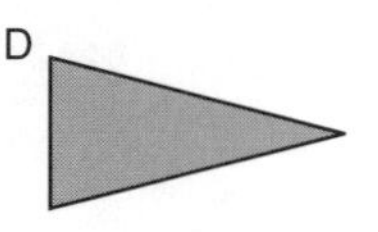

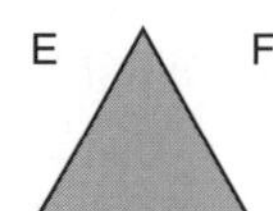

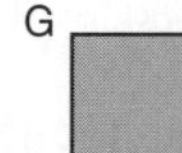

2 Copy each of these shapes onto squared paper.
Draw all the lines of symmetry on each shape.
If there are no lines of symmetry, write None.
Under each shape, write its name.

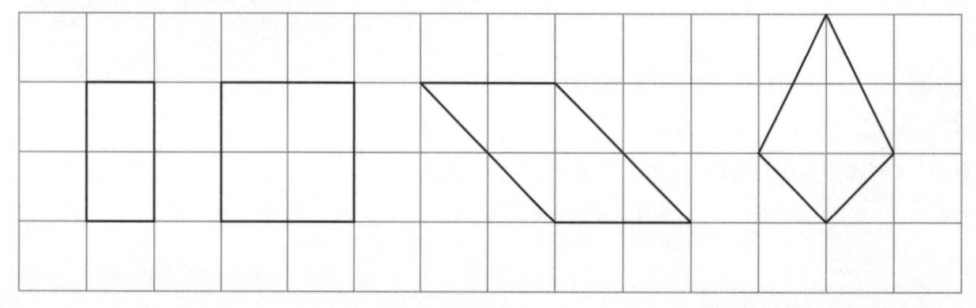

3 These are pictures of playing cards.
Do the cards have rotation symmetry?

Write *Yes* or *No* for each card.

(a) (b) (c) (d) (e)

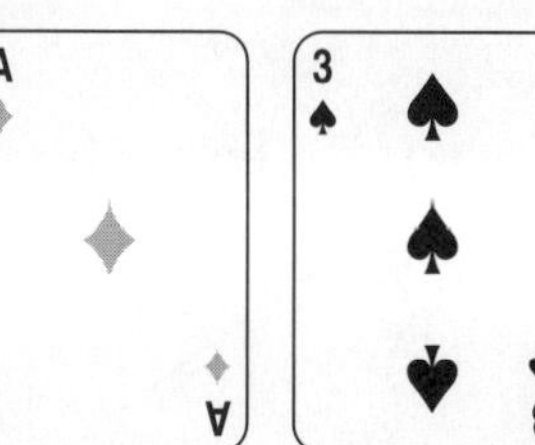

MEG (SMP)

4 How many planes of symmetry does each of these solid objects have?
If the object does not have any, write *None*.

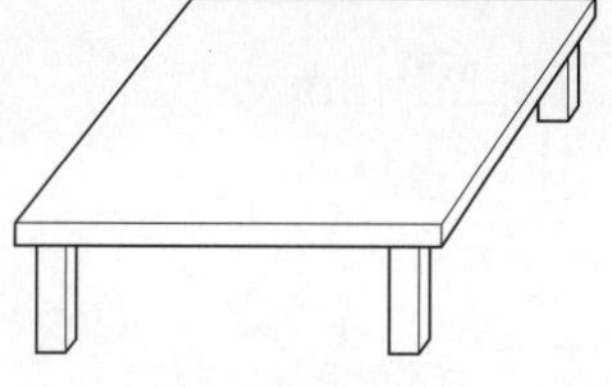
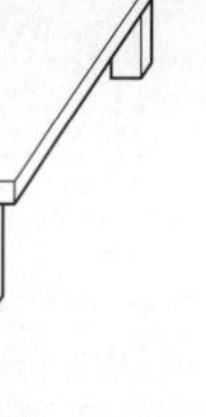

(a) Square table

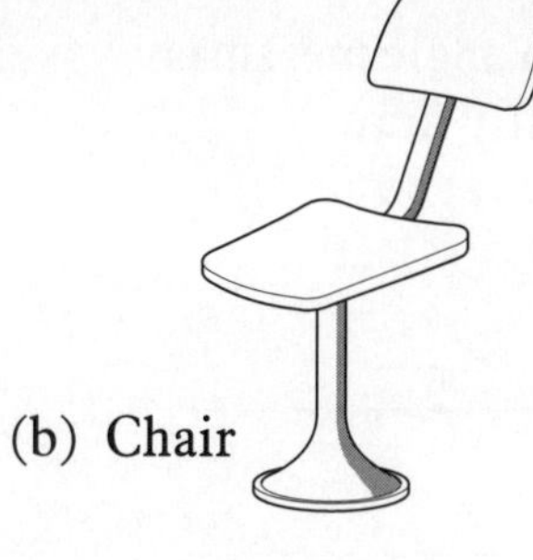

(b) Chair

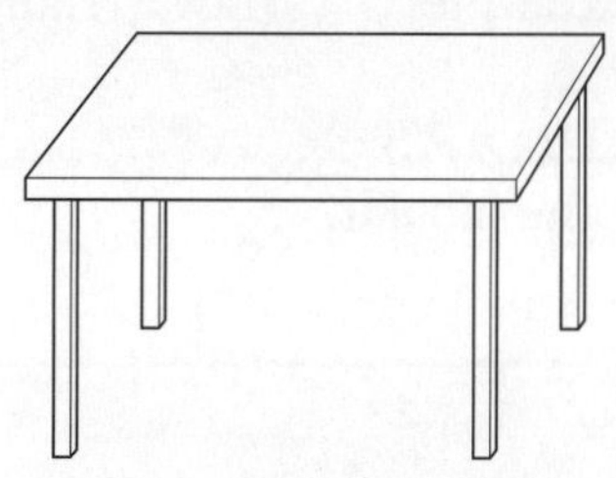

(c) Rectangular table

5 What is the order of rotation symmetry of each of these solid objects about the axis shown?

(a)

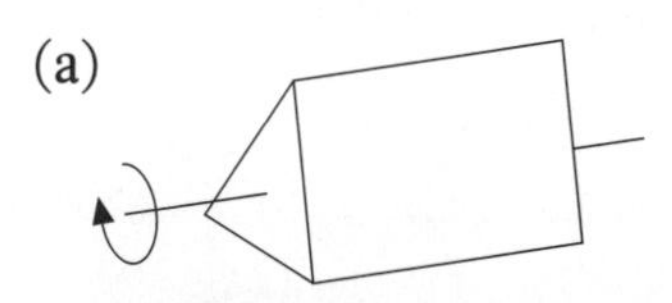

Equilateral triangular prism

(b) Square-based pyramid

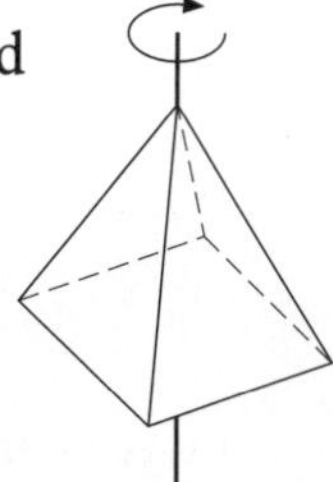

(c)

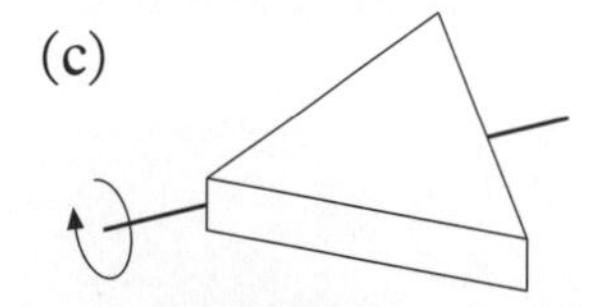

Equilateral triangular prism

Prisms ► page 49

6 Copy and complete this table. Choose the names of the quadrilaterals from:

kite, rectangle, square, trapezium, parallelogram.

Name of quadrilateral	Diagonals always cut at right-angles	Number of lines of symmetry	Order of rotation symmetry
Parallelogram			
	Yes	4	
	No	2	

Questions 7, 8 and 9 are on worksheet F4.

Constructing triangles

When you draw shapes

- Always work in pencil.
- Do not rub out your construction lines.
- It may help to draw a rough sketch first.

Example Draw a triangle with sides 3·2 cm, 5·5 cm and 4·8 cm.

Draw any side first, but it's often best to draw the longest.

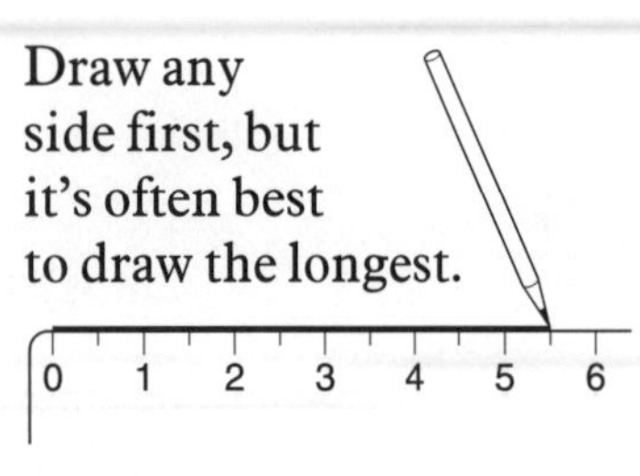

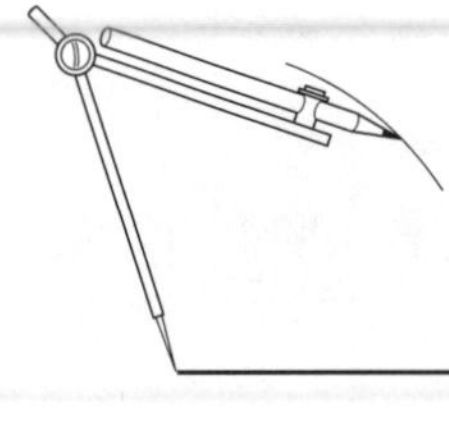

Set your compasses to 3·2cm
Draw this arc.

Draw this arc, radius 4·8 cm.

Join up the triangle.

Example Draw a triangle with AB = 4·5 cm, angle A = 38° and angle B = 65°.

Draw AB and label it.

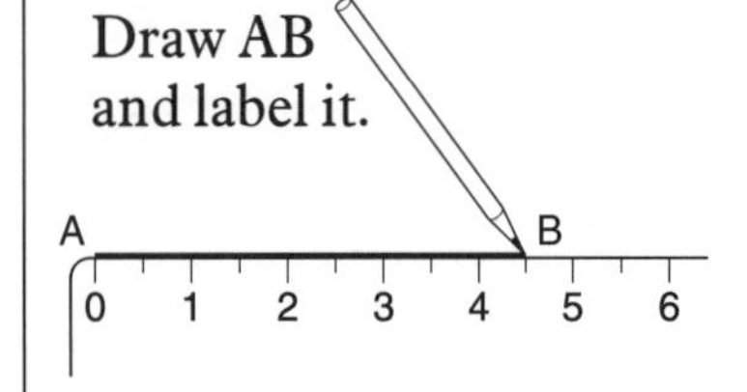

Use an angle measurer to lightly draw the side at 38°.

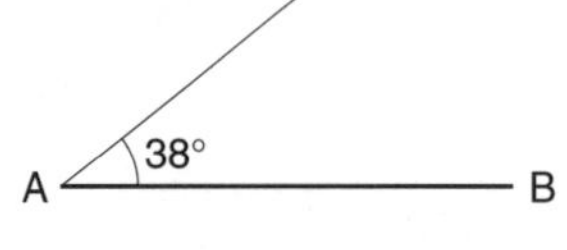

Lightly draw the 65° side.

65°
A
B

Join up the triangle.

10 Each angle at the centre of a regular hexagon is 60°.
Make a drawing of a regular hexagon.
It can be any size.

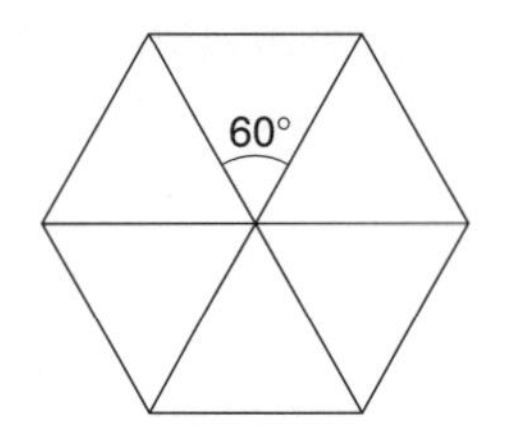

MEG (SMP)

11 This triangle has not been drawn to scale.
Make an accurate drawing of the triangle.

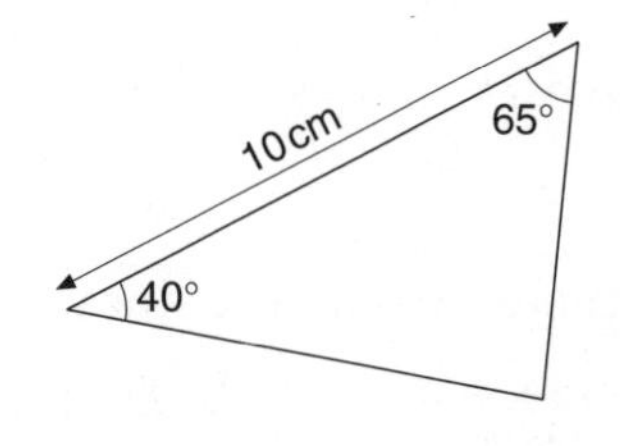

MEG/ULEAC (SMP)

12 Using only a pair of compasses, a ruler and a pencil, draw an accurate equilateral triangle.

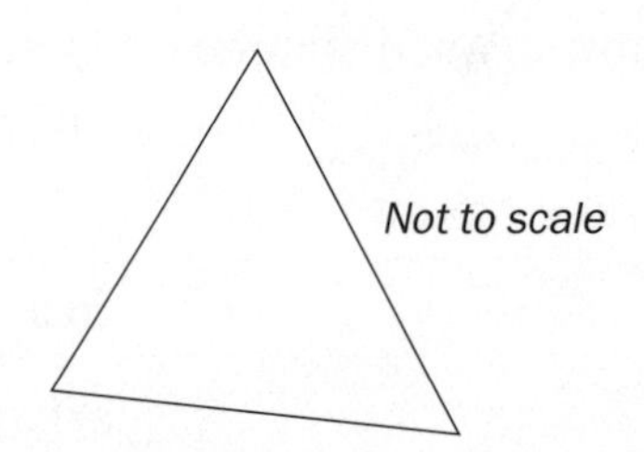

13 A design for a badge consists of three congruent isosceles triangles AEF, BFD, CDE, placed as shown.

(a) What type of triangle is DEF?

(b) Given that AE = AF = 6·2 cm and EF = 4 cm, use instruments to make a full size drawing of this design.

Congruent shapes ► page 53

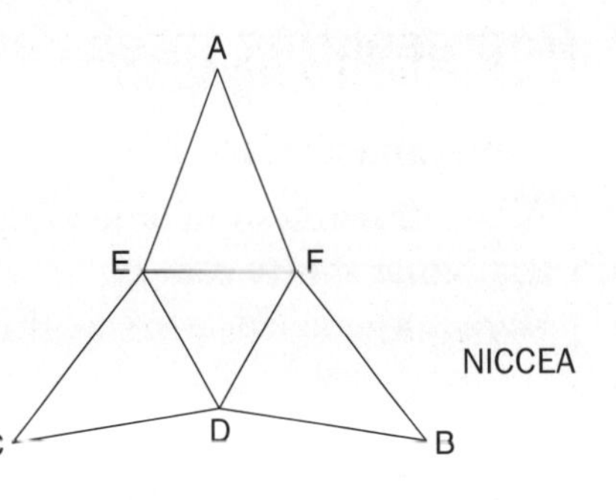

NICCEA

14 The course for a power boat race is a triangle. This is a sketch of the course, not drawn to scale.

The first leg of the course is 1000 metres.
The second leg is 1500 metres.
The angle between these two legs is 55°.

(a) Make a scale drawing of the course. Use a scale of 1 cm to 200 m.

(b) Measure the length of the third leg on your drawing.

(c) Work out the real length of this leg.

(d) What is the total length of the course?

(e) On your diagram, measure the angle between the second and third legs (marked a on the sketch).

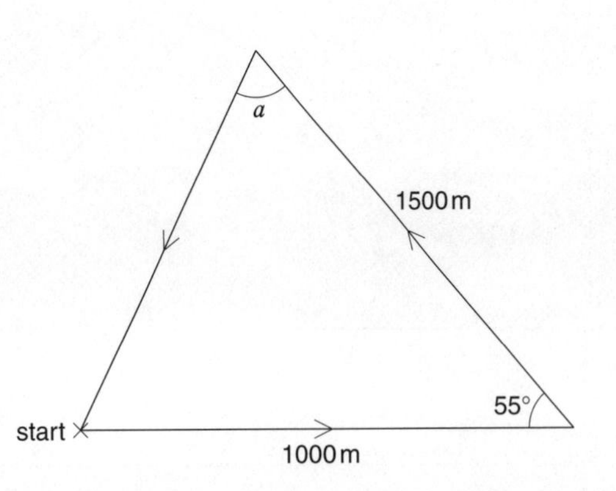

MEG/ULEAC (SMP)

15 The triangular framework shown is *not* drawn to scale.

(a) Make a scale drawing of the framework, using a scale of 1 cm to 0·1 m. Do not remove any of your construction lines.

(b) Use your scale drawing to find

(i) the size of angle BDA,

(ii) the length of AD in metres.

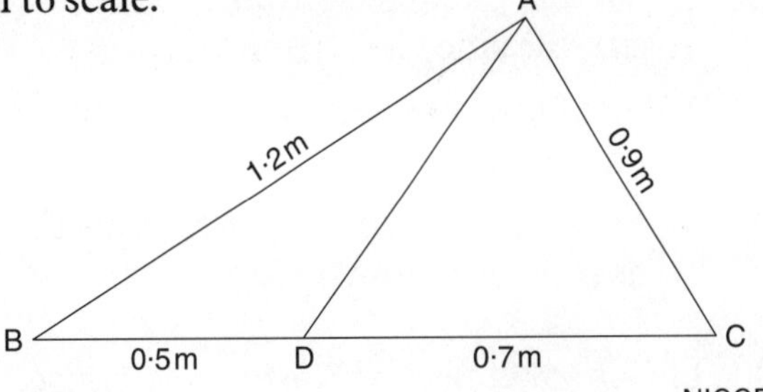

NICCEA

16 Lisa and Eve are making a pennant. This is a sketch of the pennant, not drawn accurately.

(a) Make a scale drawing of the pennant. Use a scale of 1 cm to 4 cm.

(b) Measure AB on your drawing.

(c) They are going to stick braid along the sides AB and BC of the pennant. What length of braid do they need?

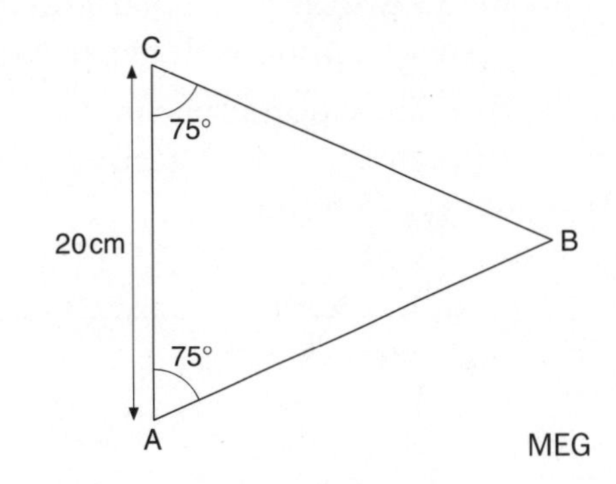

MEG

Answers and hints ► page 118

Representing three dimensions

Views and plans

When drawing solid objects, it can help to use triangular spotty paper.
Here is a garage drawn like this.

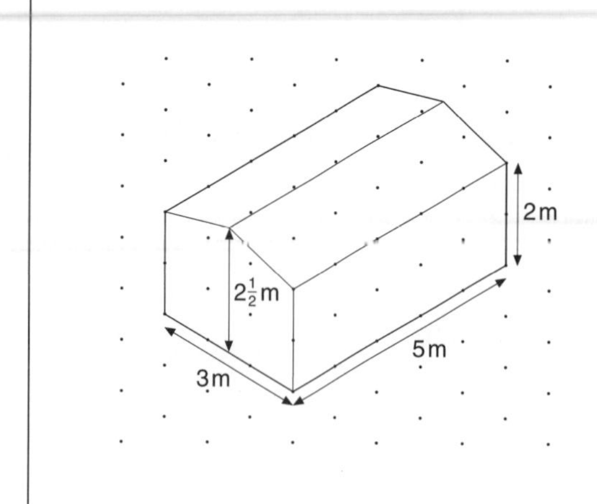

You can also draw the **plan, side view** and **front view** (or **elevation**) of the garage.
You must draw each view to the same scale.

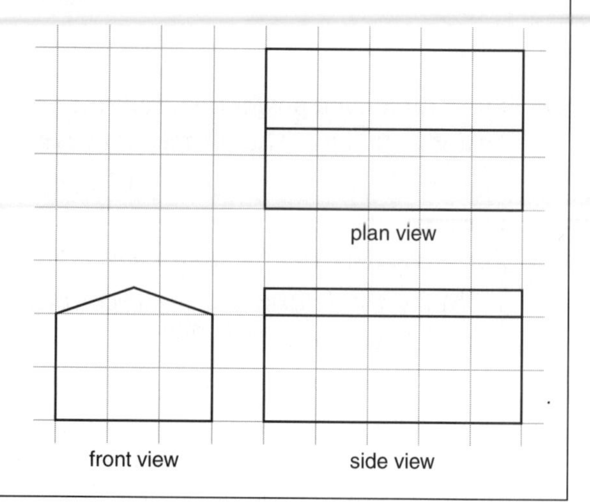

1 This is a model made from six cubes.
Each cube has edges 2 cm long.

(a) On centimetre squared paper, draw full size
  (i) the plan view (from P),
  (ii) the side view (from S).

(b) The cube marked A is taken away, and another cube is stuck to the face marked B.
On triangular spotty paper, draw the model as it appears now.

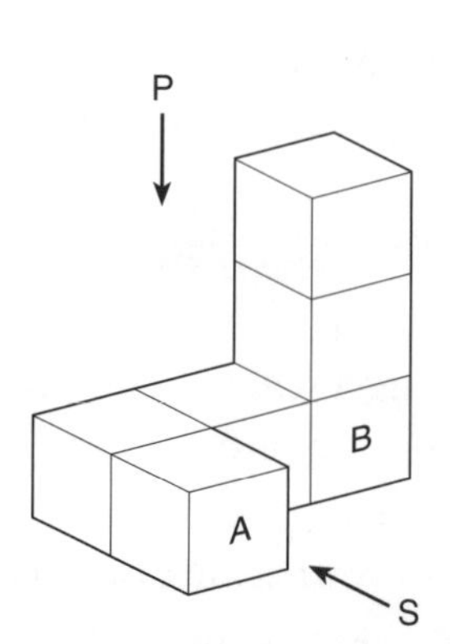

2 This is a sketch of a lean-to greenhouse.
Below is an accurate plan and front elevation of the greenhouse, drawn to a scale of 1 cm to 1 m.

By taking appropriate measurements from the plan and elevation, draw accurately the side elevation of the greenhouse.

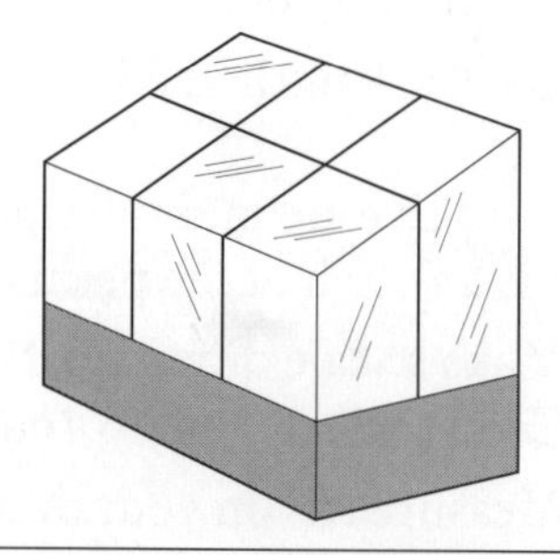

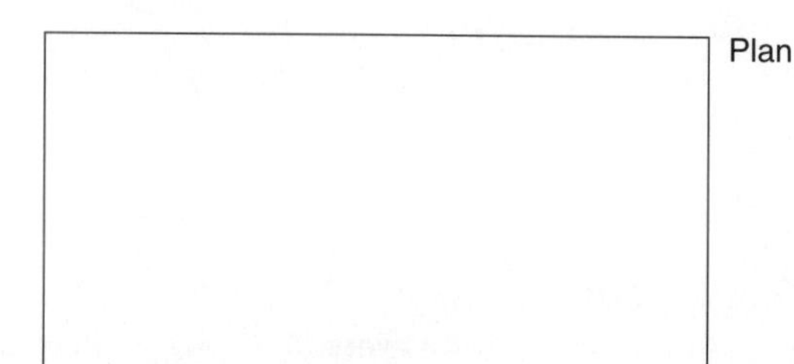

Nets

The **net** of a three-dimensional object is a shape you can cut out and fold up to make the object.
This is the net of a **prism**, drawn full-size.

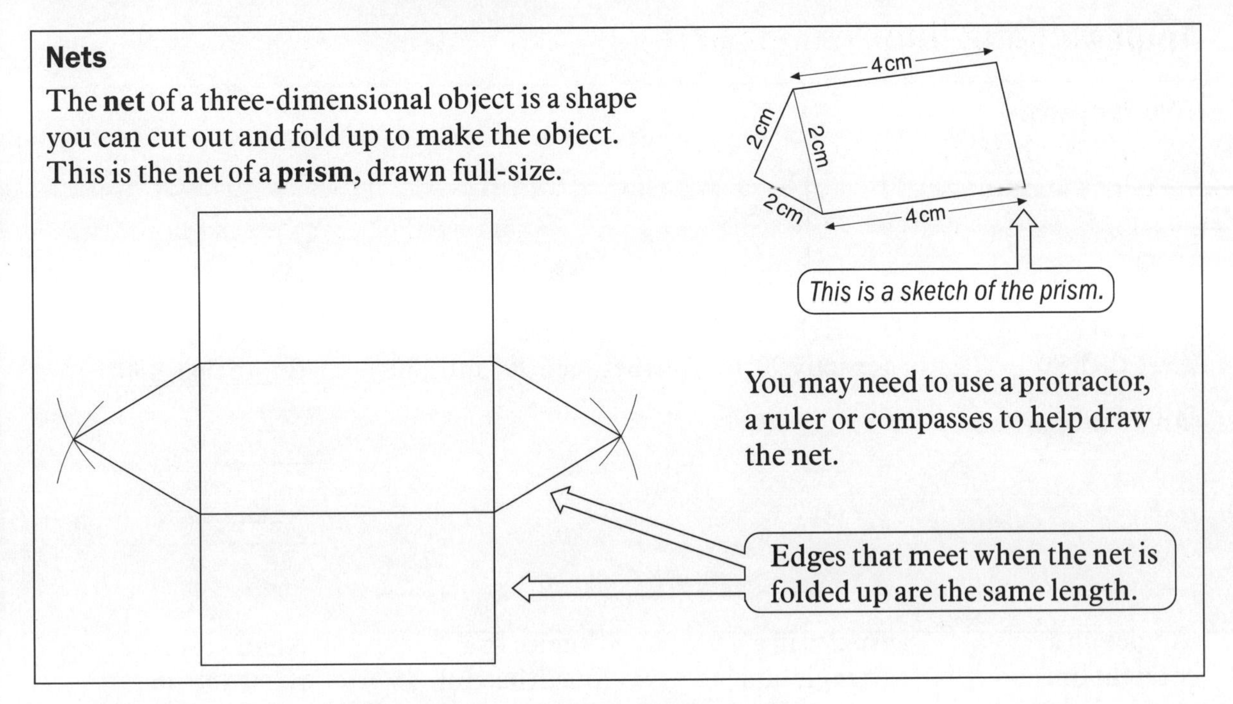

3 Here are some nets. Some of them fold up to make a cube.
Write down the letters of the nets that fold to make cubes.

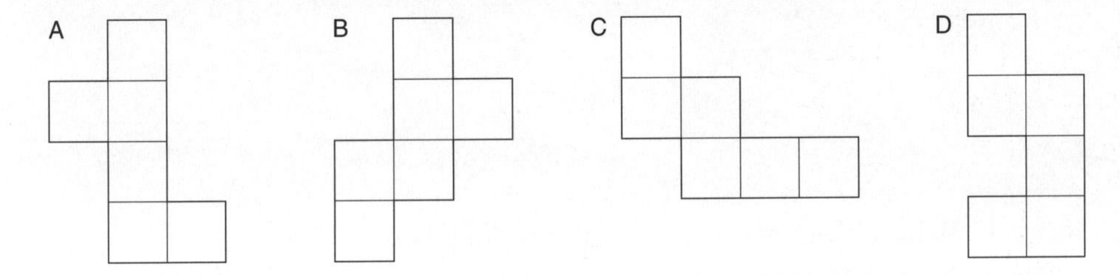

4 These diagrams show a packing case in two positions.

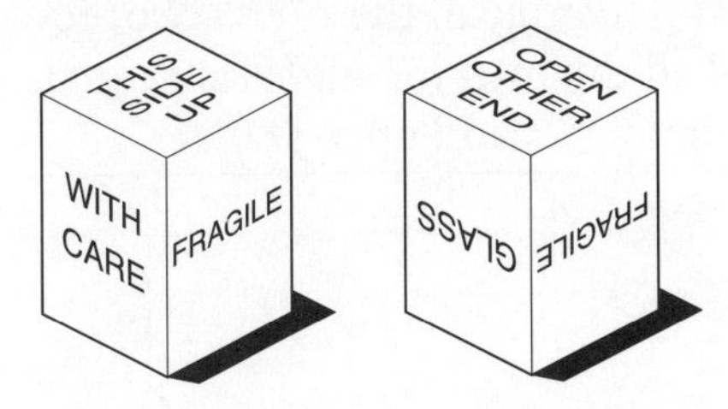

Copy this sketch of the net of the packing case and complete the labelling on the net.

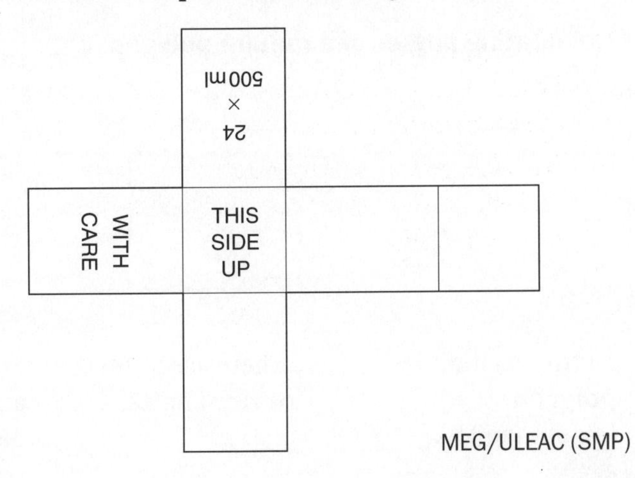

MEG/ULEAC (SMP)

Questions 5 to 8 are on worksheets F5 to F8.

Answers and hints ► page 121

Angles

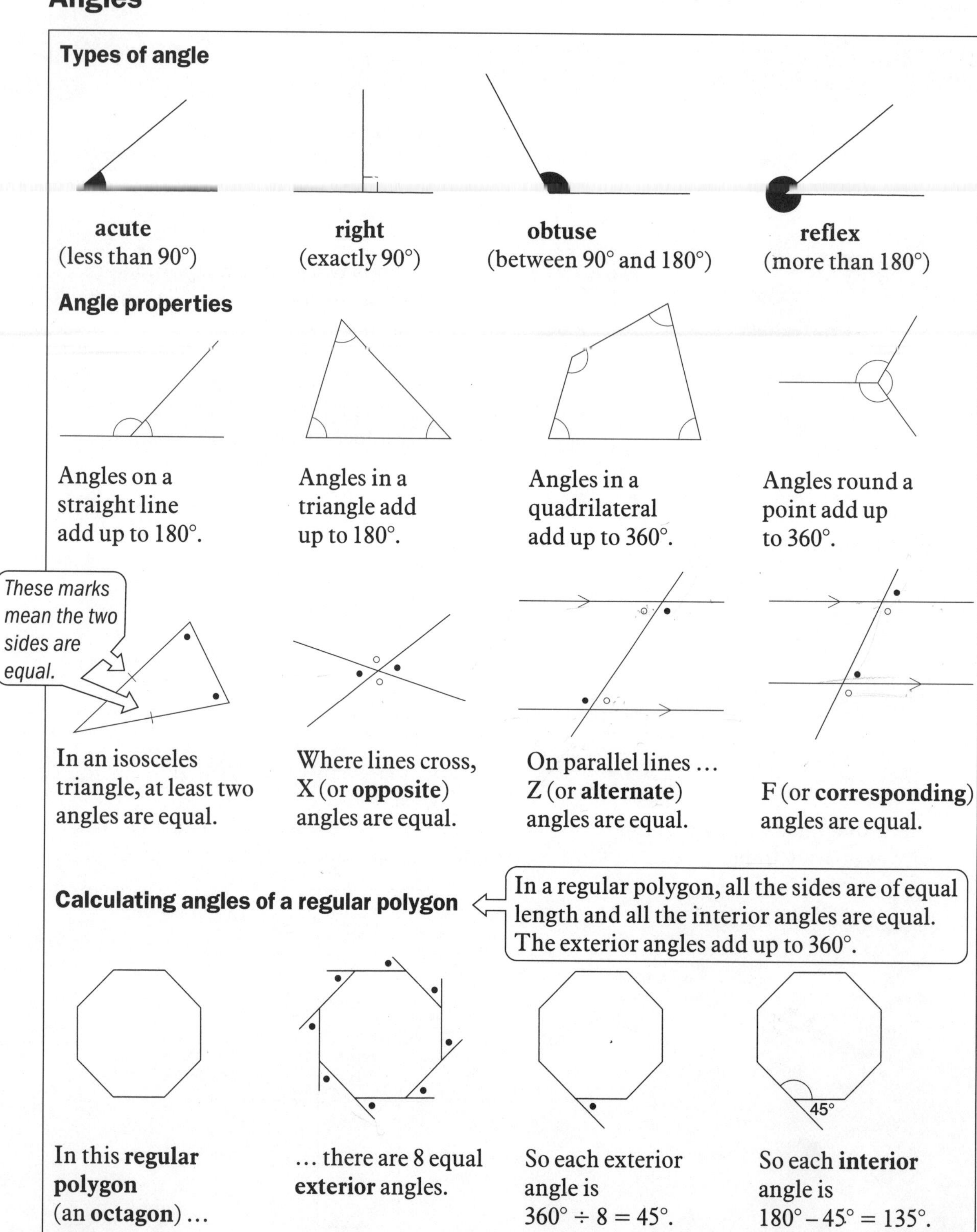
Types of angle
acute
(less than 90°)
right
(exactly 90°)
obtuse
(between 90° and 180°)
reflex
(more than 180°)
Angle properties
Angles on a straight line add up to 180°.
Angles in a triangle add up to 180°.
Angles in a quadrilateral add up to 360°.
Angles round a point add up to 360°.
These marks mean the two sides are equal.
In an isosceles triangle, at least two angles are equal.
Where lines cross, X (or opposite) angles are equal.
On parallel lines …
Z (or alternate) angles are equal.
F (or corresponding) angles are equal.
Calculating angles of a regular polygon
In a regular polygon, all the sides are of equal length and all the interior angles are equal. The exterior angles add up to 360°.
45°
In this regular polygon (an octagon) …
… there are 8 equal exterior angles.
So each exterior angle is 360° ÷ 8 = 45°.
So each interior angle is 180° – 45° = 135°.
Polygon names: A pentagon has 5 sides, a hexagon 6 sides and an octagon 8 sides.

None of the diagrams is to scale.

1 Work out the values of the angles marked with letters.

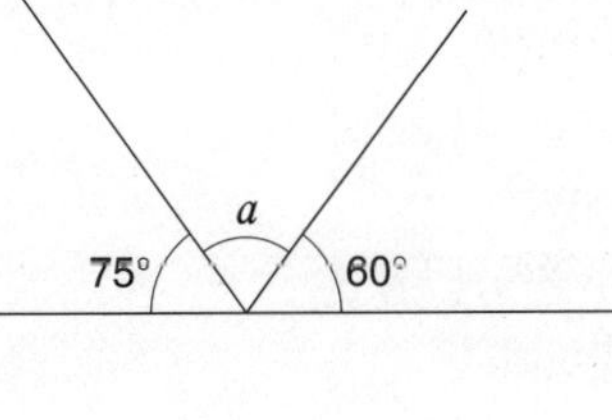

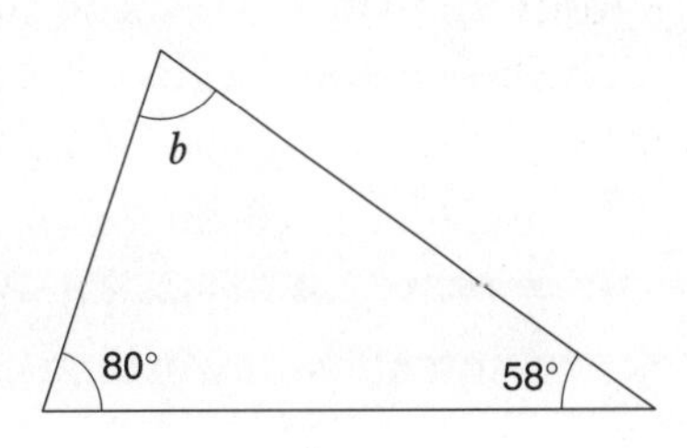

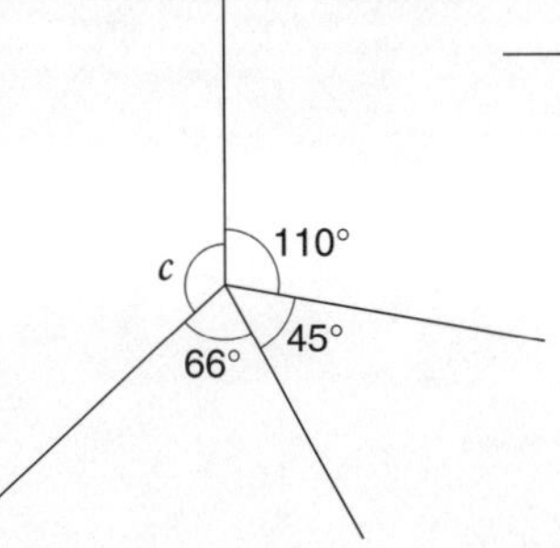

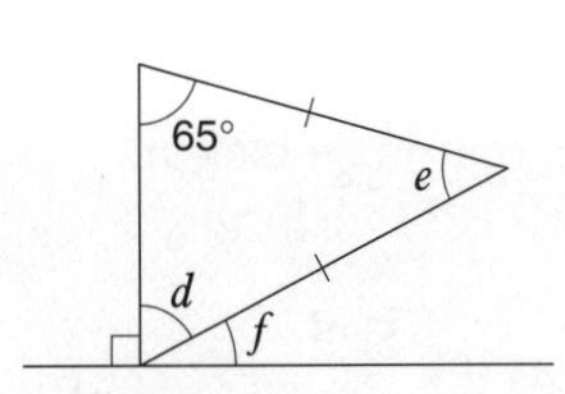

2 Find the sizes of the angles marked by letters in these diagrams.

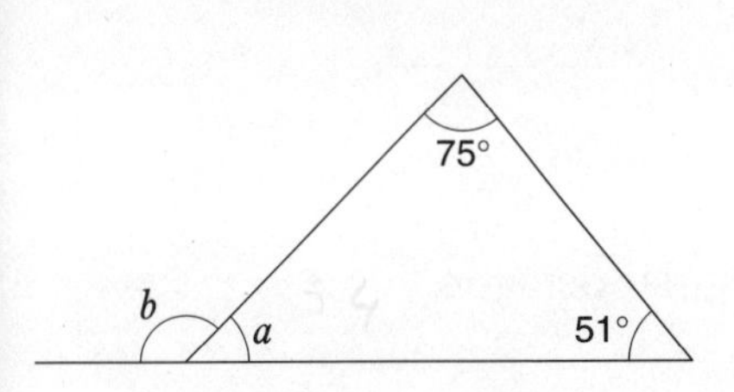

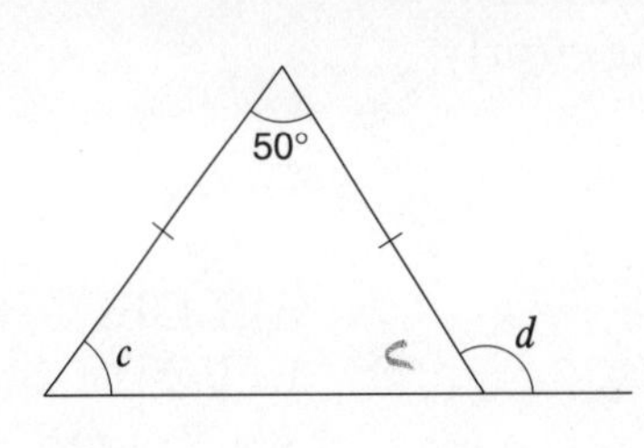

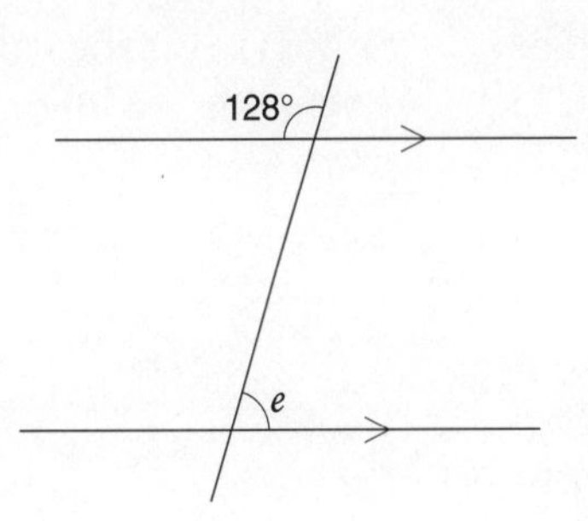

3 Find the angles marked p, q, r, s and t. Show all your working and give reasons for your answers.

For example:

$p = 45°$, because the angles of a triangle add up to 180°.

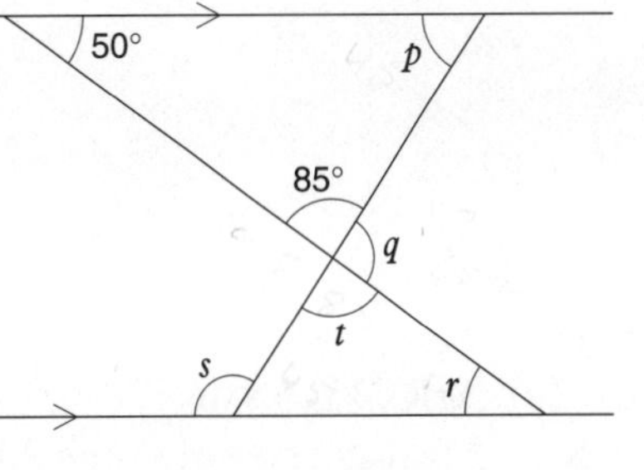

4 (a) Calculate the values of:

(i) p

(ii) q

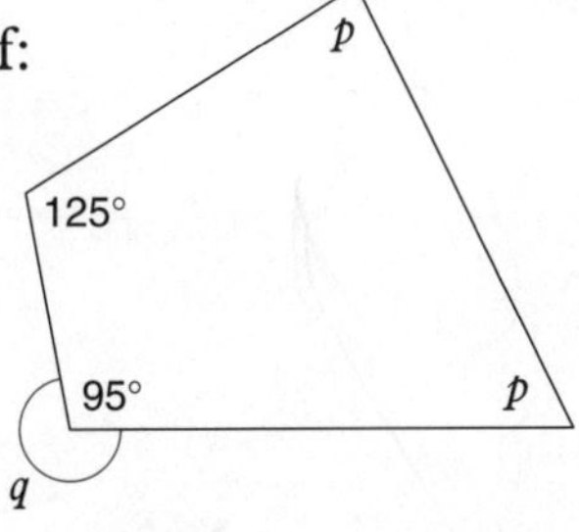

(b) Calculate the values of:

(i) w

(ii) x

(iii) y

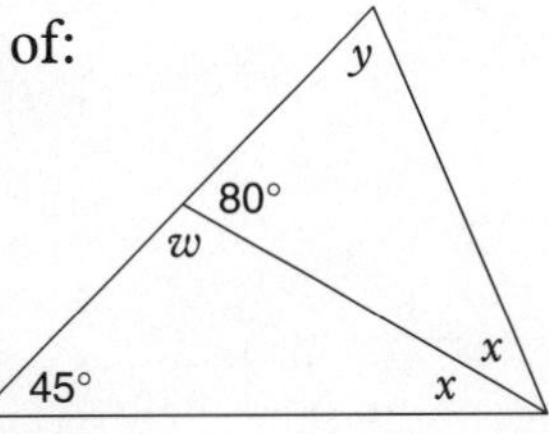

NICCEA

5 An ironing board rests on two legs of equal length when open. The board and the floor are horizontal.

(a) Calculate the size of the angle marked x and give a reason for your answer.

(b) Calculate the size of the angle marked y. Show all your working clearly.

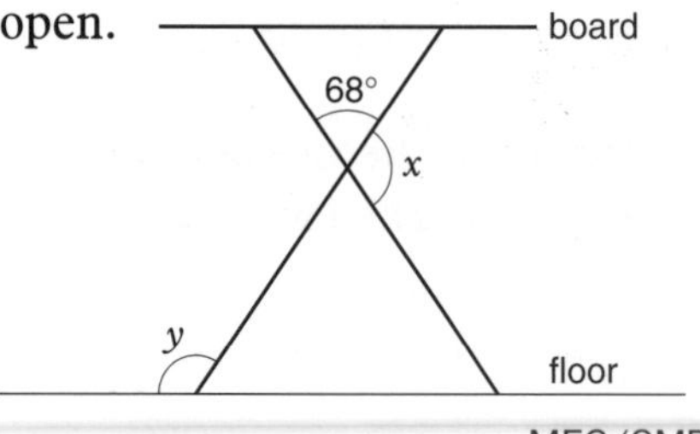

MEG (SMP)

6 In the diagram, ABCDEF is a regular hexagon. Triangle EPD is isosceles, with PE = PD.

(a) Calculate angle r. Give reasons for your answer.

(b) Calculate the angle marked s, giving reasons.

(c) Work out the value of angle t, showing your working clearly.

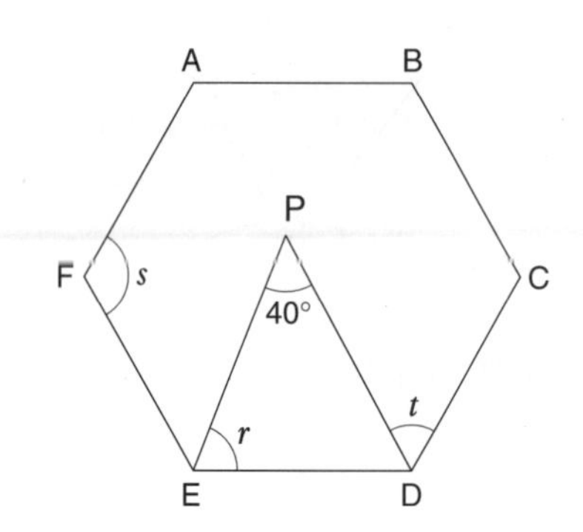

7

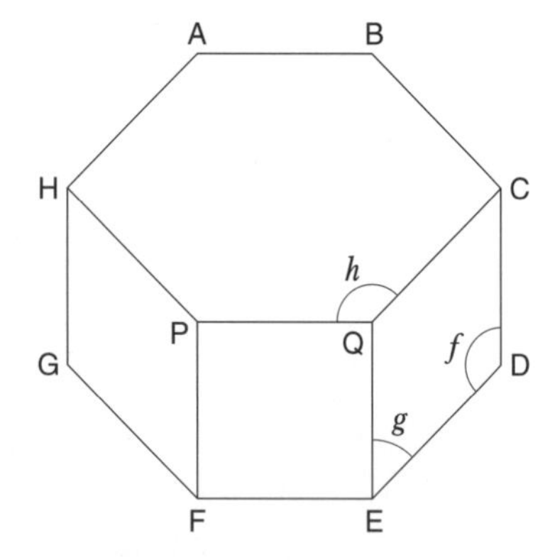

ABCDEFGH is a regular octagon. EFPQ is a square. Work out the values of the angles marked f, g and h.

8 (a) ABCD is a kite. Find the size of angle ABC.

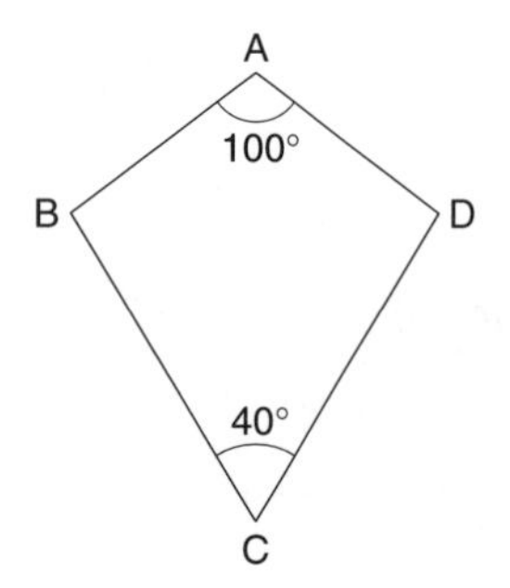

(b) In the figure, TS = TP, SP is parallel to RQ and SR is parallel to PQ.

Find the values of:

(i) x

(ii) y

(iii) z

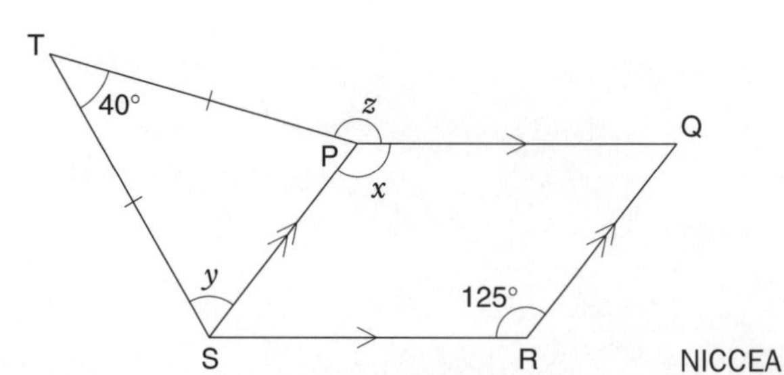

NICCEA

Answers and hints ► page 123

Transformations

Translation, rotation and enlargement

When you are drawing rotations or reflections it is a very good idea to use tracing paper. You should be able to ask for this, even if you are not given a piece at the start of your examination.

This is a **translation** of A to B.
The translation is **3 across** and **1 up**.

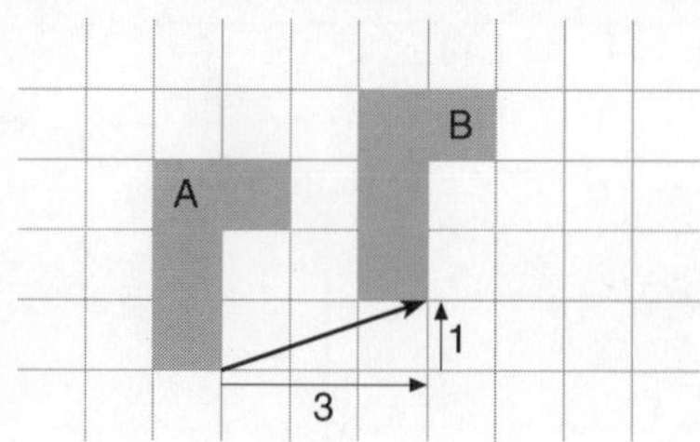

This is a **reflection** in the mirror line shown. You can draw reflections by tracing. Then fold the tracing paper along the mirror line.

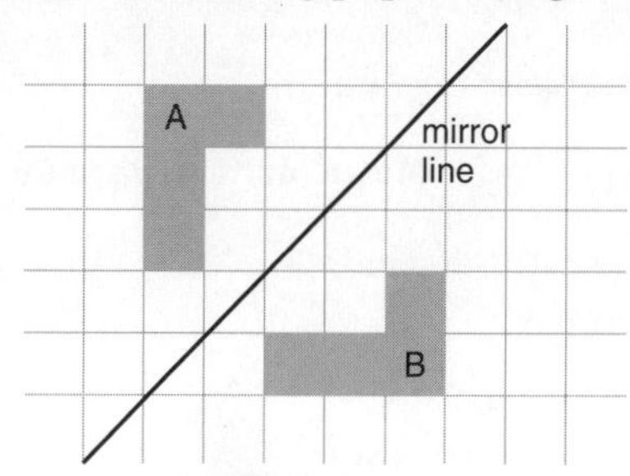

This is a **rotation** of $\frac{1}{4}$**-turn** (or **90°**) **anti-clockwise** with **centre** P.

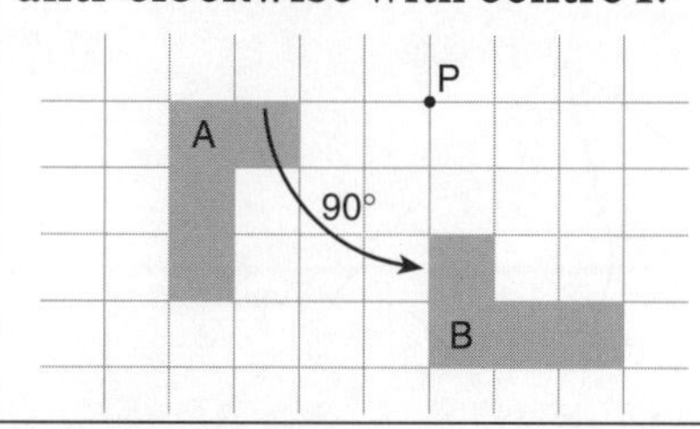

Shapes that are the same shape and size are **congruent** to each other.
Here A, C and D are congruent.

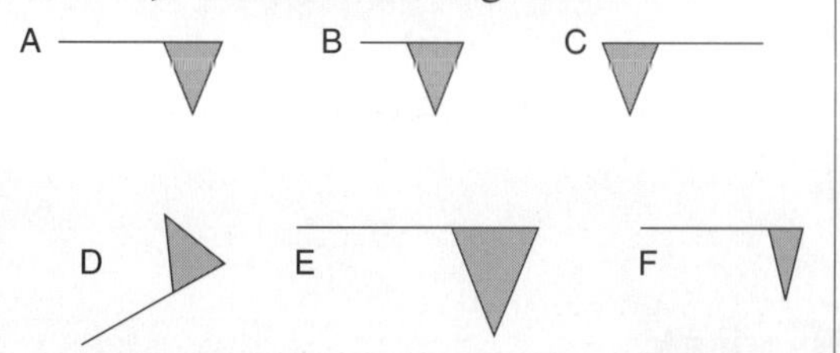

B is an **enlargement** of A with **scale factor 2**.
The lengths of B are twice as long as A.

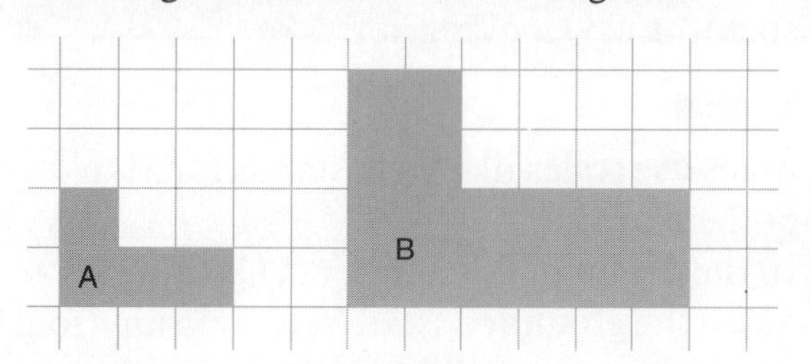

You may have to enlarge from a centre.
Measure from the centre to the first point.
Multiply the distance by the scale factor.
Measure this distance **from the centre** again.

If the scale factor is 2,
OA′ is twice OA.

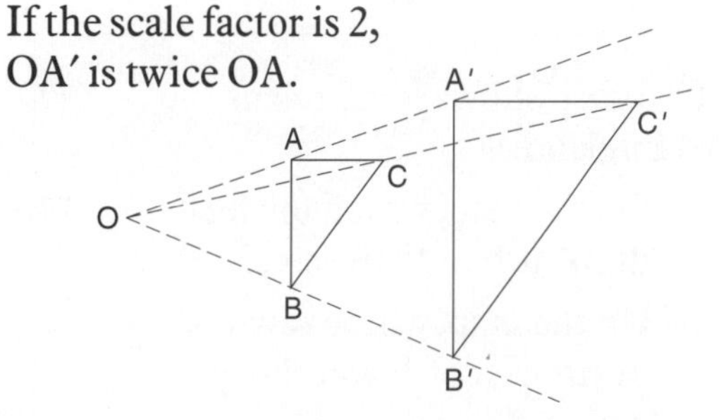

Tessellations

These **tessellations** cover the page with pieces that are the same shape. There are no gaps.

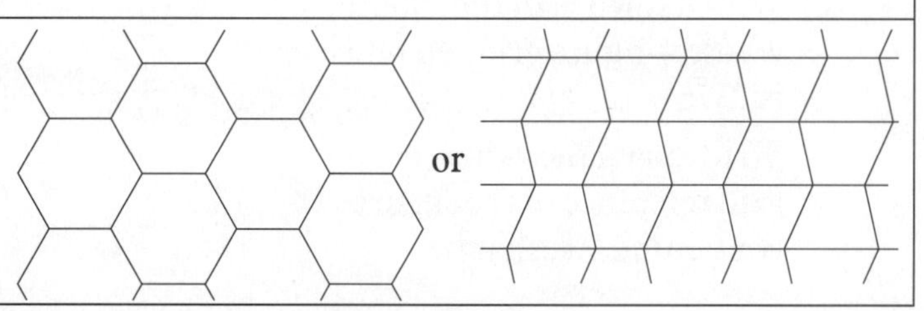

All the questions on transformations are on worksheets F9 to F12.

Answers and hints ► page 123

Scales and bearings

Grid references

The point B is in a square on this map.
The coordinates of the bottom left-hand corner of the square are (13, 25).

So we say that B is in the square with **4-figure grid reference** 1325.

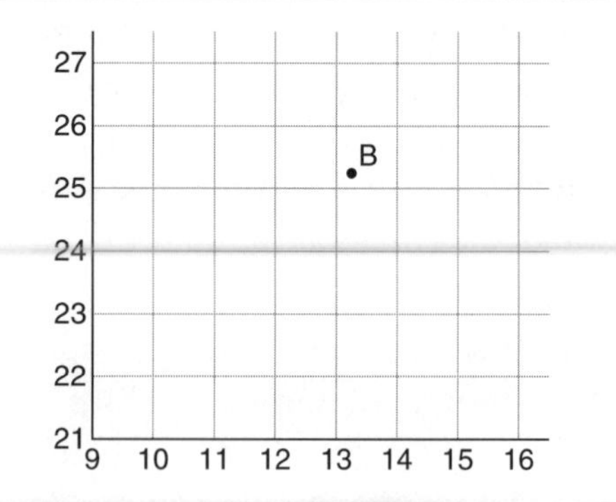

Bearings

Bearings are measured **clockwise** from North.
They are given in degrees as 3-figure numbers.

'The bearing of P from Q' means you are standing at Q looking at P.

Scales

Maps use scales like '1 cm stands for 5 km', or '1 cm to 5 km'.
In the diagram, P is 3 cm from Q.
So on the ground, P is 3×5 km = 15 km from Q.

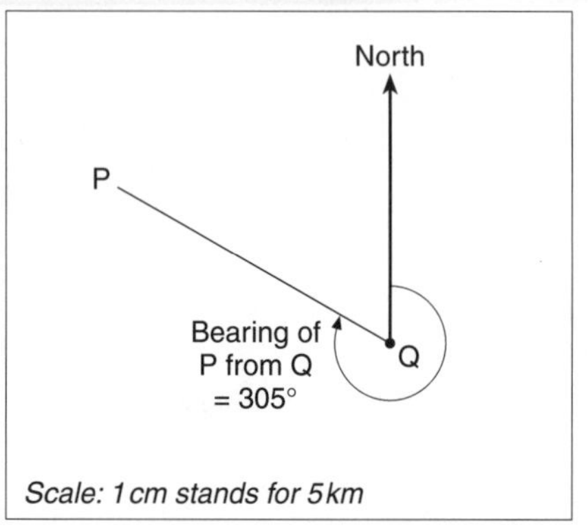

Scale: 1 cm stands for 5 km

Metric units ► page 56

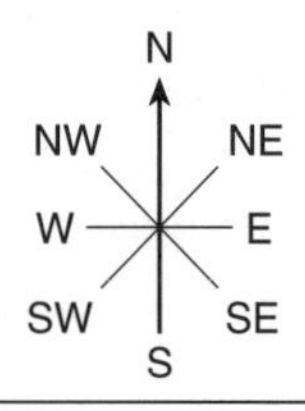

Points of the compass

Check that you know the 8 points of the compass:
N (North), NE (North-East), E (East), SE (South-East), S (South), SW (South-West), W (West), NW (North-West).

1 This map shows some towns in England.

(a) On the map, which town is due south of Worcester?

(b) On the map, which town is north-east of Tewkesbury?

(c) Which town is in the square whose 4-figure grid reference is 2376?

(d) Write down the 4-figure grid reference of the square containing Evesham.

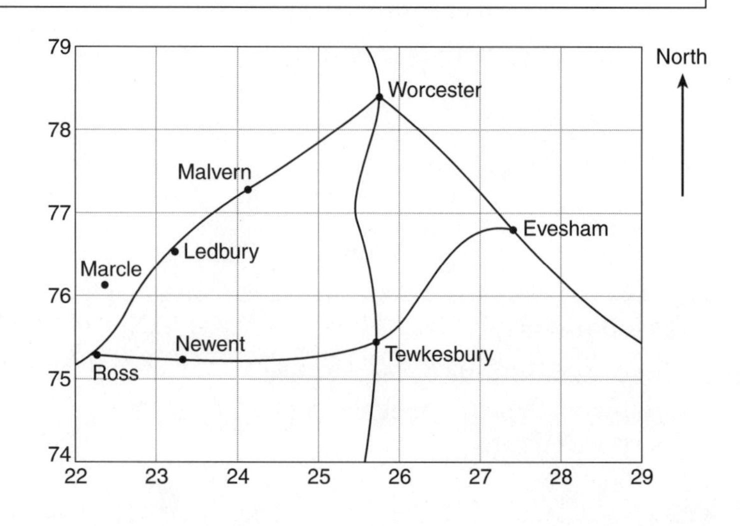

2 This is a street plan of a housing estate in Abingdon.

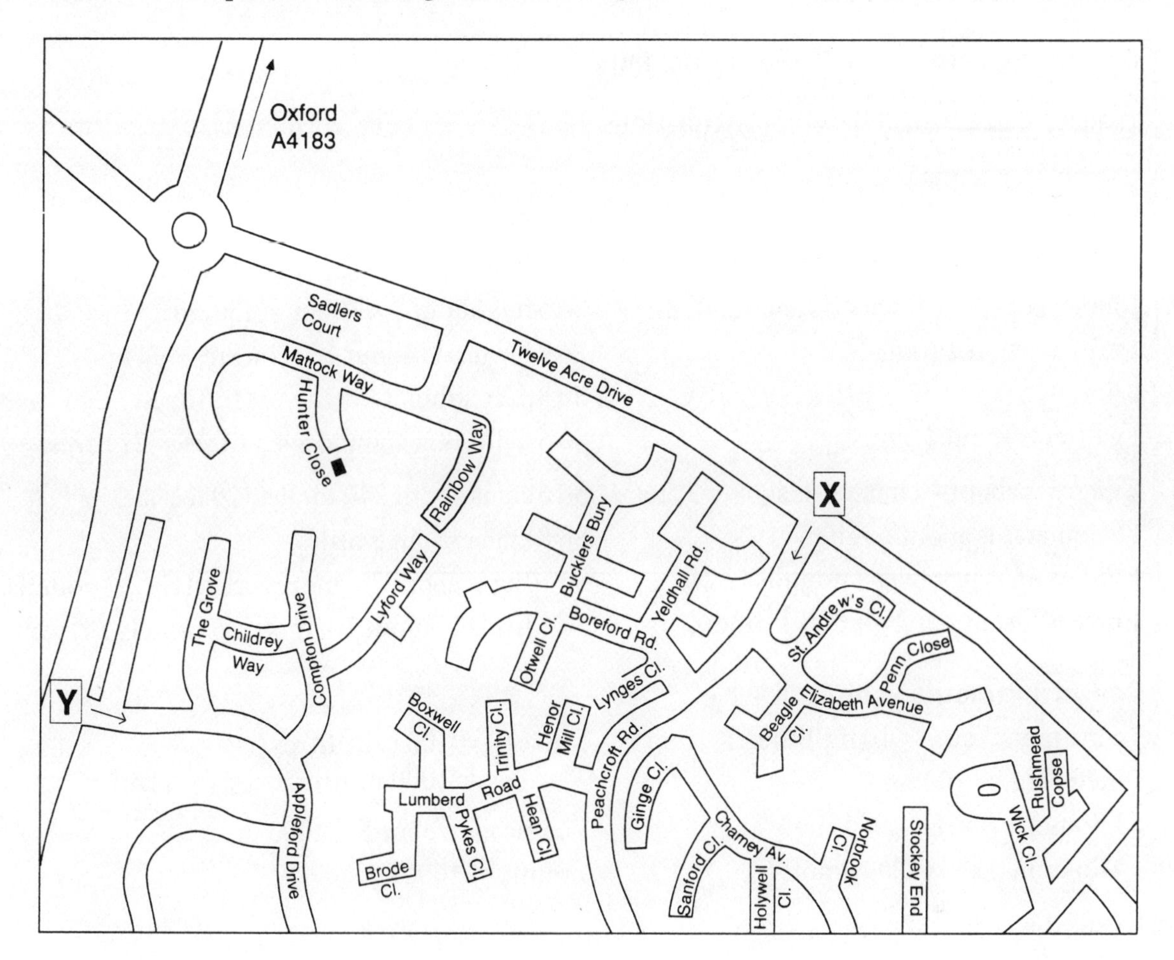

(a) A motorist enters the housing estate at **X**.
He takes the first turn on the left and then the second on the left.
Where is he?

(b) A jogger enters the estate at **Y**.
She takes the second turn to the left into The Grove.
She then turns right, right again and immediately left.
Where is she now?

(c) A cyclist lives in Hunter Close.
His house is shown by a ■ on the map.
Copy and complete the following simple instructions
to tell him how to get to the A4183 road to Oxford.
Come out of Hunter Close, turn right, then ...

MEG (SMP)

Questions 3 to 5 are on worksheets F13 to F15.

Answers and hints ► page 125

Units and measures

Converting between 'old' and metric units

You need to know all these approximations.

1 pound (lb) is about $\frac{1}{2}$ kg. So 6 lb is about $6 \times \frac{1}{2}$ kg = 3 kg.

1 inch (in or ") is about 2·5 cm or 25 mm. So 4 in is about $4 \times 2{\cdot}5$ cm = 10 cm or 100 mm.

1 foot (ft or ' = 12 inches) is about 30 cm. So 5 ft is about 5×30 cm = 150 cm.

1 mile is about 1·6 km. So 15 miles is about $15 \times 1{\cdot}6$ km = 24 km.

1 pint (pt) is just over $\frac{1}{2}$ litre. So 3 pt is about $3 \times \frac{1}{2}$ litre = $1\frac{1}{2}$ litres.

1 gallon is about 4·5 litres. So 6 gallons is about $6 \times 4{\cdot}5$ litres = 27 litres.

1 metre is about 40 inches (just over 3 feet). So 6 cm is about 240 inches (20 feet).

1 kilometre is about $\frac{5}{8}$ mile. So 8 km is about 5 miles.

1 kilogram is just over 2 pounds (2·2 lb). So 10 kg is about 20 (more accurately 22) pounds.

1 litre is just under 2 pints ($1\frac{3}{4}$ pints). So 8 litres is about 16 (more accurately 14) pints.

Converting metric units

1 centimetre (cm) = 10 millimeters (mm)

1 metre (m) = 100 cm

1 kilometre (km) = 1000 m

1 kilogram (kg) = 1000 grams (g)

1 litre (l) = 100 centilitres (cl)
= 1000 millilitres (ml) (1 ml = 1 cm^3)

1 cubic metre (m^3) = 1000 litres (l)

1 tonne = 1000 kg

Examples

36 cm = 36×10 mm = 360 mm

$0{\cdot}6\,m^3 = 0{\cdot}6 \times 1000$ litres = 600 litres

48 mm = $48 \div 10$ cm = 4·8 cm

2400 litres = $2400 \div 1000\,m^3 = 2{\cdot}4\,m^3$

1 (a) Roughly how many pints of milk does this container hold?

(b) Roughly how many kilograms does this bag of rice hold?

MEG (SMP)

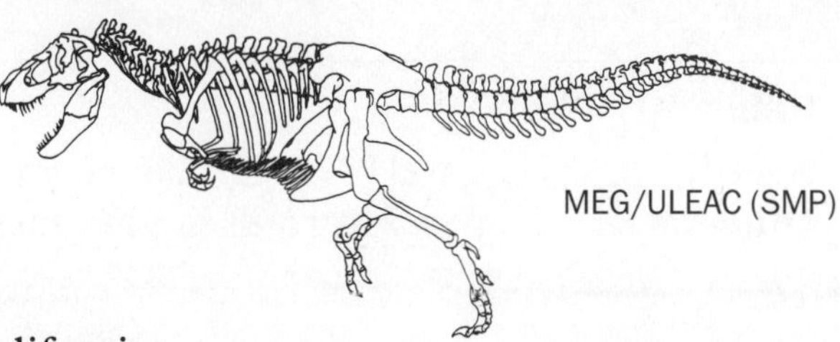

2 One of the largest fossils of a flesh-eating dinosaur is 20 feet tall.
About how many metres tall is this? MEG/ULEAC (SMP)

3 The north-west face of the Half Dome in California is 2200 feet high.
It is almost vertical and is the world's steepest mountain climb.
Approximately how high is the north-west face in metres? MEG/ULEAC (SMP)

4 Work out the rough metric equivalents of the following.

(a) The weight in kilograms of a dog that weighs 40 pounds.
(b) The length in millimetres of a worm that is 2 inches long.
(c) The volume in litres of a petrol tank that holds 8 gallons.
(d) The distance in kilometres from London to Edinburgh, 400 miles.

5 In each of these examples, say which is greater.
Show your working clearly.

(a) 5 litres or 3 pints (b) 100 inches or 1 metre
(c) 100 millimetres or 1 foot (d) 100 grams or 1 pound

6 Change each of these to the units stated.

(a) 15 000 metres to kilometres (b) 0·5 metres to centimetres
(c) 150 millilitres to litres (d) 1·5 kilograms to grams

7 Alan cycles 1·6 km on his bike and then walks 600 metres.
How far does he travel altogether?

8 

A standard measure of rum is 25 ml.
How many standard measures can be served from a 1 litre bottle? MEG (SMP)

9 The total weight of a pile of sheets of paper is 4 kg.
Each sheet of paper weighs 5 g.
How many sheets are in the pile? MEG (SMP)

10 Rosalyn pours 300 ml of orange squash and 1·5 litres of water into a jug.
How many litres of liquid are in the jug altogether? MEG (SMP)

Rates

We say kilometres per hour.

Speed: If I travel 80 km in 5 hours, my average speed is 80 km ÷ 5 hours = 16 km/h.

Other rates: If I type 240 words in 8 minutes, my typing rate is 240 words ÷ 8 minutes = 30 words per minute.

Time graphs ► page 32

Estimating

Unless a question says otherwise, it is useful to remember that:

A fairly tall person is about 1·8 metres or 180 centimetres tall.
A double-decker bus is about 4·5 metres high.
A desk is about 75 centimetres high.

Reading scales

Make sure you can read the number each arrow points to. (The answers are at the bottom of the page.)

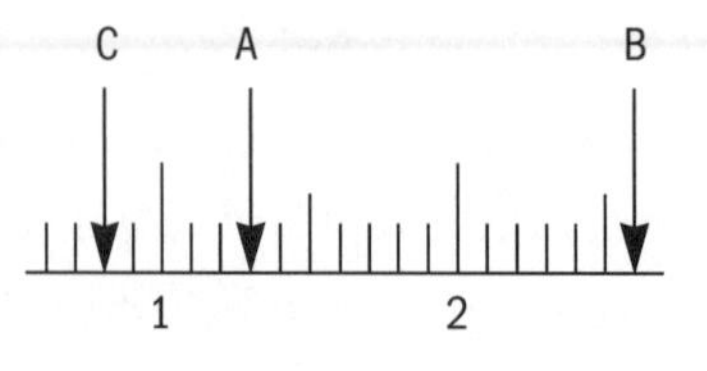

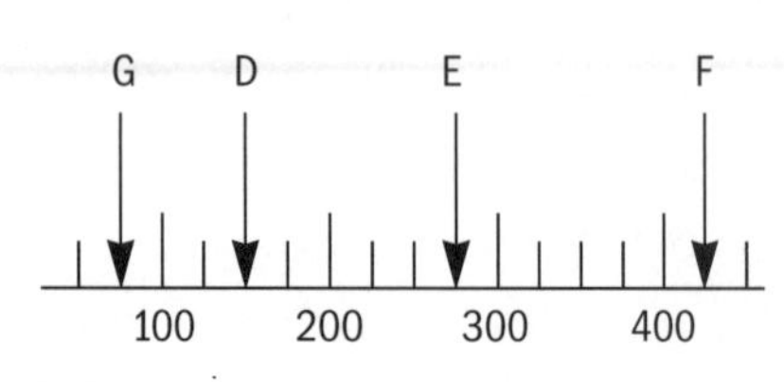

11 In an airport, luggage is carried on a conveyor belt.
The belt moves at 2·5 metres per second.

(a) How long does it take for a case to travel 30 metres?

(b) Kamaljit's luggage takes 18 seconds to reach him from the start of the belt. How far has it travelled?

MEG/ULEAC (SMP)

12 One lap of a race track is 5 km.
A car covers this distance in 2 minutes.

What is the average speed of the car in kilometres per hour for this lap?

NICCEA

13 Tony and Carol own a narrow boat.
They took 16 minutes to travel through the Foulridge Tunnel.
The tunnel is 1500 metres long.
Work out their average speed in metres per **second**.

MEG/ULEAC (SMP)

14 Work out the heart-rate for each of these patients in beats per minute.

(a) Paul's heart beat 150 times in 3 minutes.

(b) Sue's heart beat 20 times in 15 seconds.

(c) Liz's heart beat 90 times in 45 seconds.

15 Spen takes exactly 1 minute to run 400 metres.

(a) What is his speed in metres per **second**?

(b) After the race his heart beats 45 times in 20 seconds.
What is his heart rate in beats per minute?

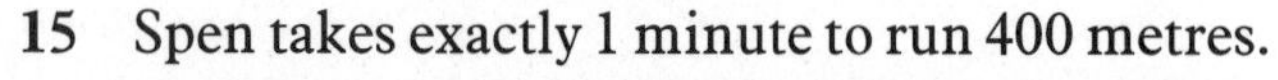

Questions 16 to 20 are on worksheets F16 and F17.

A 1·3, B 2·6, C 0·8, D 150, E 275, F 425, G 75

Answers and hints ► page 127

Length, area and volume

The **perimeter** of a shape is the distance round the outside of it.

Perimeter = 5 m + 3 m + 5 m + 3 m
= 16 m

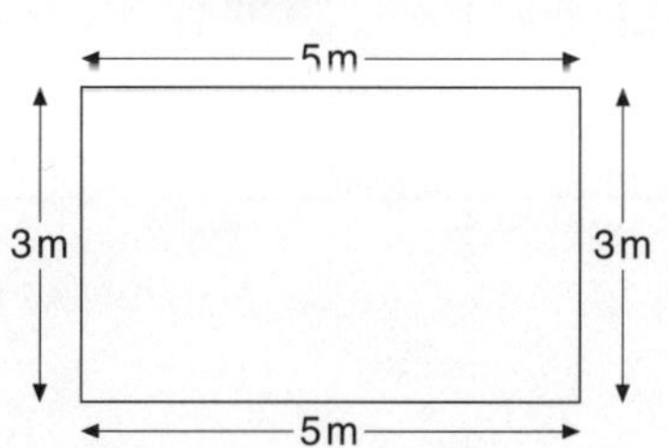

The **area** of a rectangle = length × width.
We measure area in units like cm^2 or m^2.

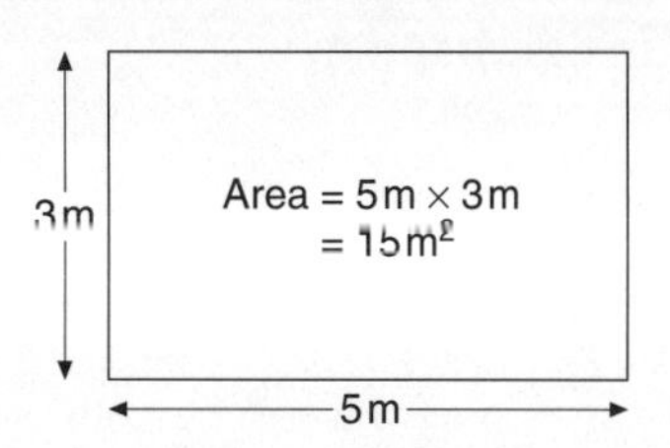

A **cuboid** is a shape like a shoe box.
You can count centimetre cubes to find the volume of a cuboid.

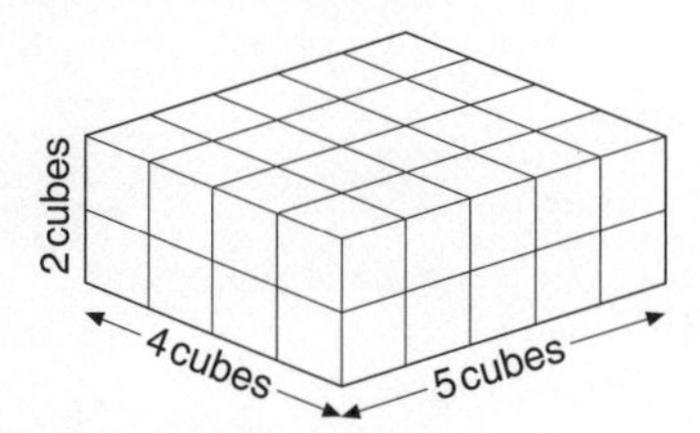

There are 5 × 4 × 2 = 40 centimetre cubes here.

Volume of a cuboid
= length × width × height
We measure volume in units like cm^3 or m^3.

Volume
= 5 cm × 4 cm × 2 cm
= 40 cm^3

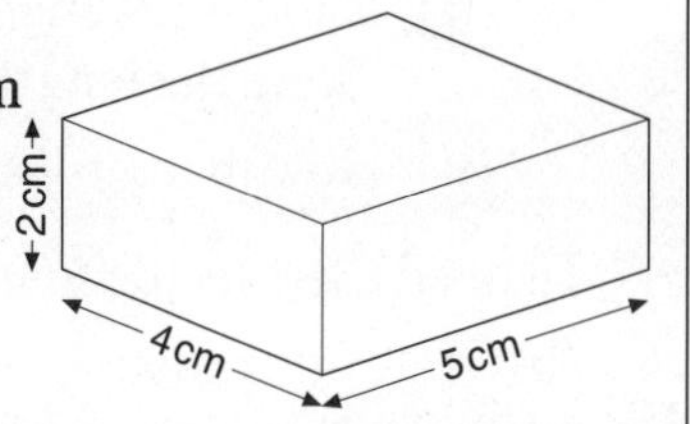

Litres ► page 56

1 This is a centimetre square grid.

(a) Find the area and perimeter of the rectangle.

(b) On centimetre square paper draw another rectangle which has the same area as this one, but with a different perimeter.
What is the perimeter of your rectangle?

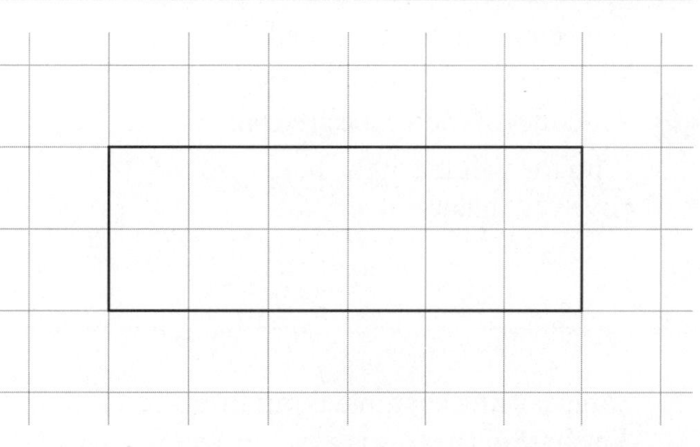

MEG/ULEAC (SMP)

2

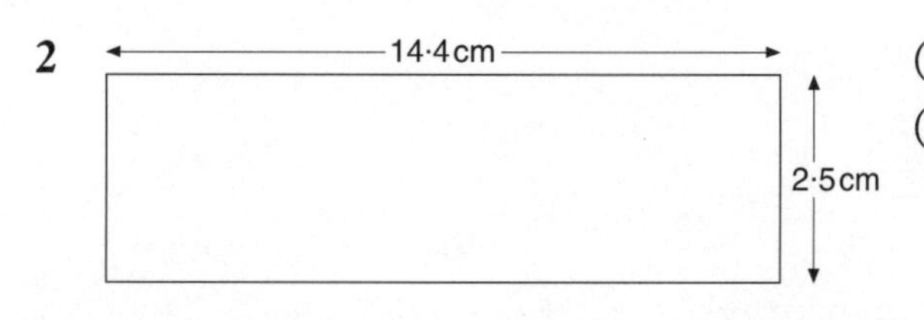

(a) Calculate the area of this rectangle.

(b) (i) What is the length of each side of the square which has the same area as the rectangle?

(ii) Draw this square accurately. MEG

3 Look at shape A.

(a) What is the area of shape A?

(b) On centimetre squared paper, draw a rectangle with the same area as shape A.

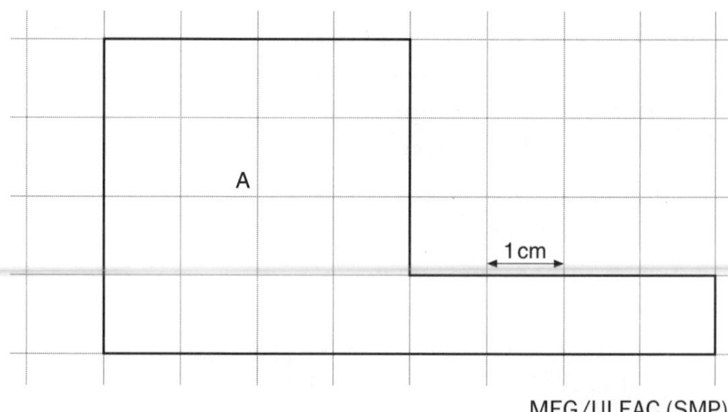

MEG/ULEAC (SMP)

4 Members of Southlands Youth Group meet in a large hall measuring 14 m by 10 m.

They were given 60 carpet tiles, each 1 metre square, to carpet part of the hall as a lounge area.

They wanted to make a square with some of the tiles. They did not cut any of the tiles.

(a) (i) What is the biggest square they could make? (State the length of the side.)

(ii) How many tiles would this use?

Instead they decide to use all the tiles and make a rectangle.

(b) State one possible rectangle they could make that would fit in the hall, using all the tiles.

MEG/ULEAC (SMP)

5 A rectangular box is packed full of sugar lumps. The sugar lumps are all cubes of equal size.

The sides of each cube are 1 cm.

Find the volume of the box. Give your answer in cm^3.

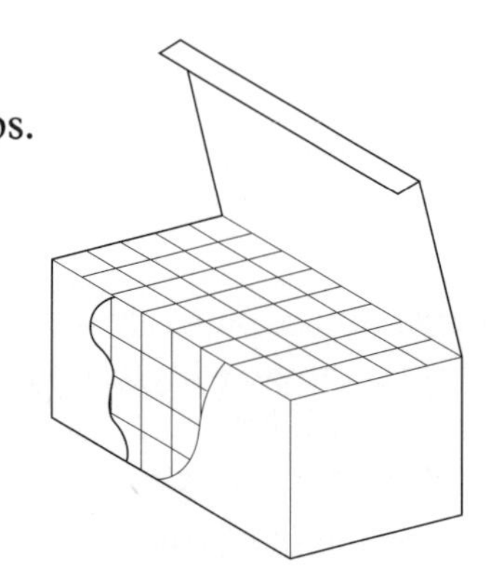

MEG/ULEAC (SMP)

6 Some potting compost is put into a tray. The base of the tray is a rectangle 35 cm by 22 cm. The depth of the potting compost is 5 cm.

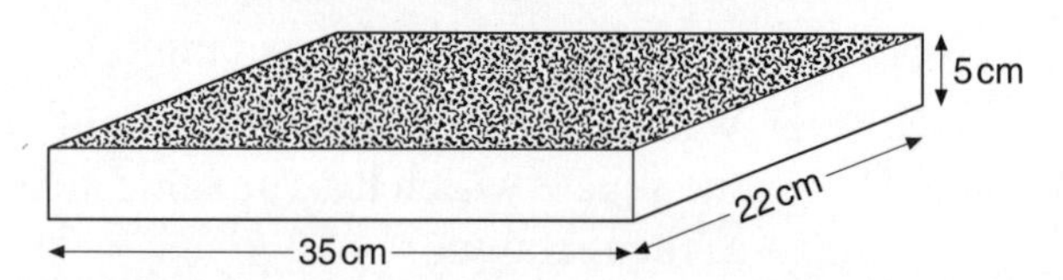

What is the volume of the potting compost in the tray?

MEG (SMP)

7 This is a mini-pack of biscuits.

(a) Work out the volume of the mini-pack.

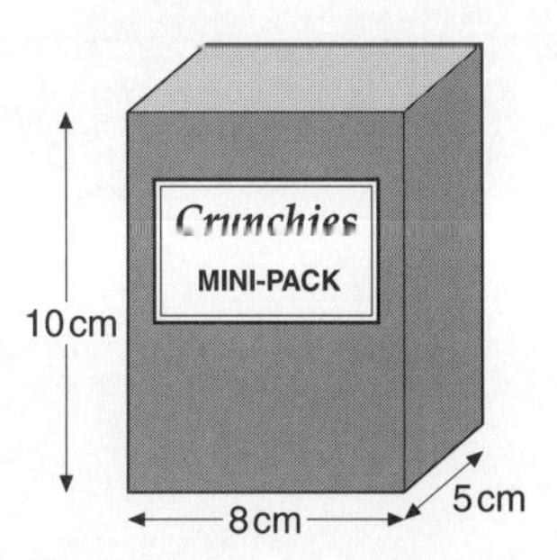

(b) A multi-pack box measures 24 cm by 10 cm by 20 cm.

How many mini-packs can be fitted into the multi-pack box?

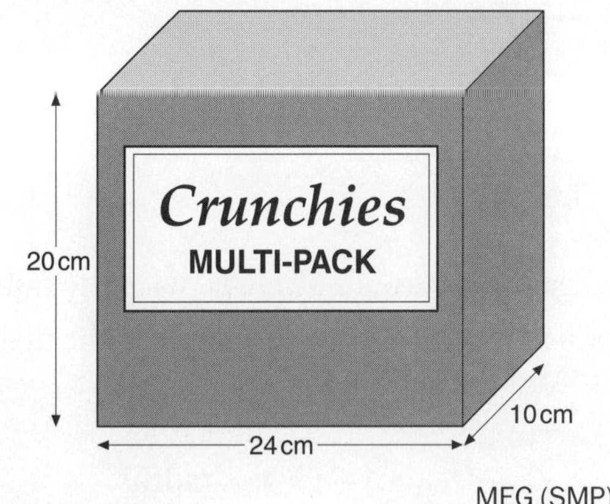

MEG (SMP)

8 This pack of coffee is a cuboid.

(a) Calculate the volume of the pack.

(b) (i) Calculate the area of the top of the coffee pack.

(ii) Calculate the total surface area of the six faces of the coffee pack.

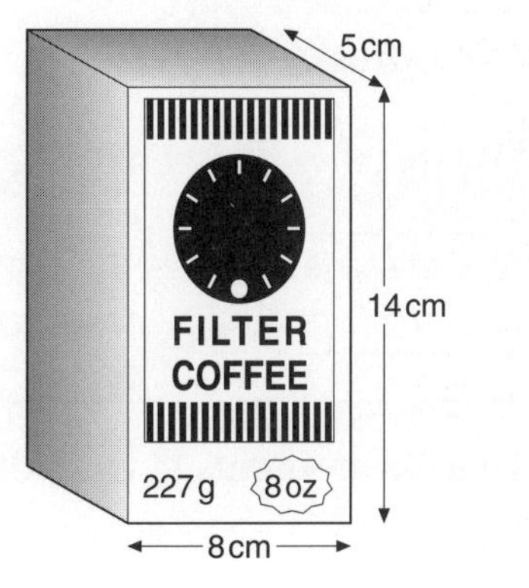

MEG (SMP)

9 This is a fish tank.
The fish tank is made from five pieces of glass.
(There is no glass at the top.)

(a) What is the area of the front piece of glass?

(b) What is the total area of glass?

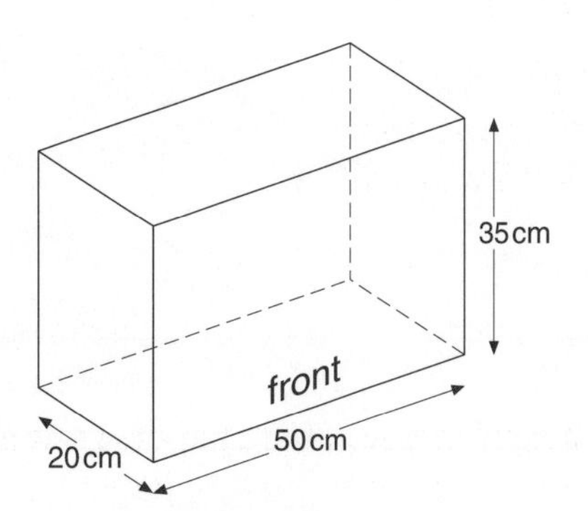

Water is put into the tank until the level of the water is 5 cm below the top.

(c) What is the volume of the water in cm^3?

(d) Cathy wants to add 'Aquafresh' to the water.
She needs to add 1 ml of Aquafresh for each litre of water in the tank.
How much Aquafresh should she add?

Answers and hints ► page 128

Triangles and circles

The area of a triangle = $\frac{1}{2}$ base × height

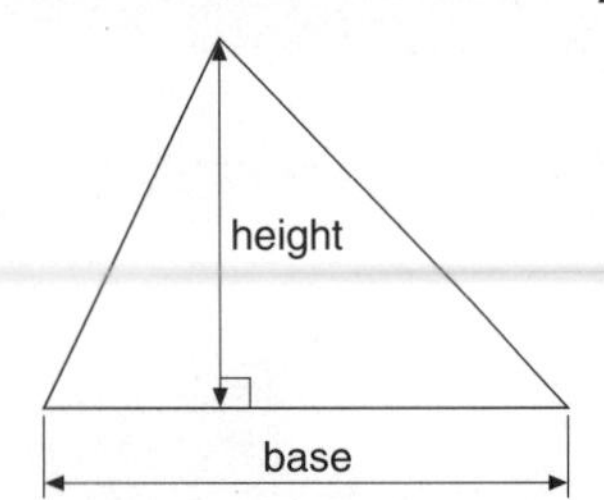

The base can be any side.

The height does not have to be vertical.

Just measure it at 90° to the side you choose as the base.

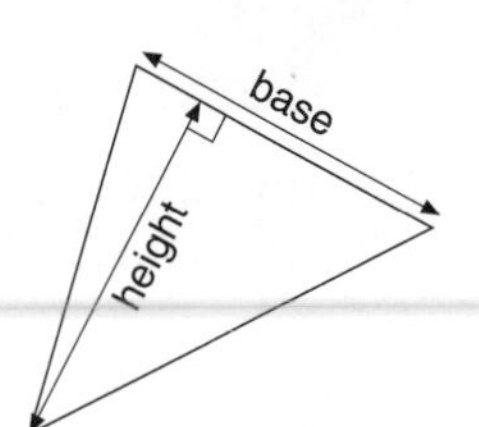

The **diameter** of a circle is twice the radius.

The **circumference** of a circle is the distance round the outside.

Circumference $= \pi \times$ diameter
$= 2 \times \pi \times$ radius

The **area** of a circle
$= \pi \times (\text{radius}) \times (\text{radius})$
$= \pi \times (\text{radius})^2$

If your calculator doesn't have a π key, use the value 3·14.

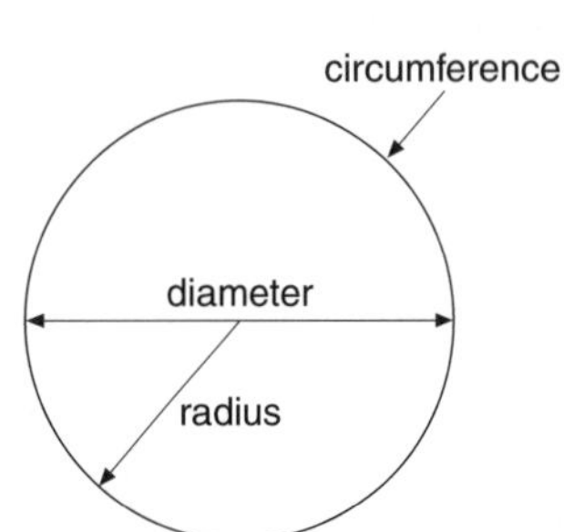

You also need to know these names.

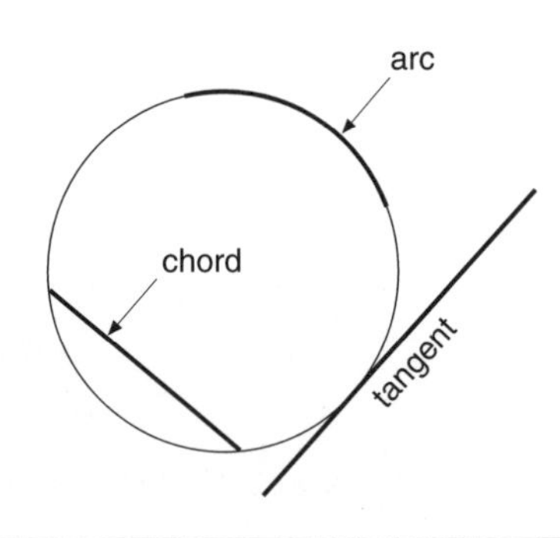

1 Take your own measurements to work out the area of each of these.

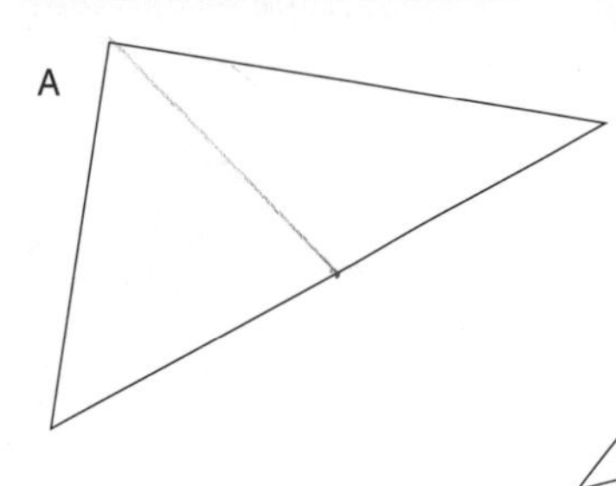

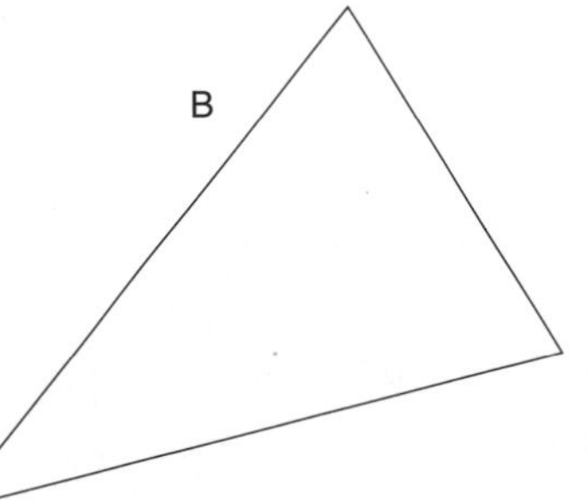

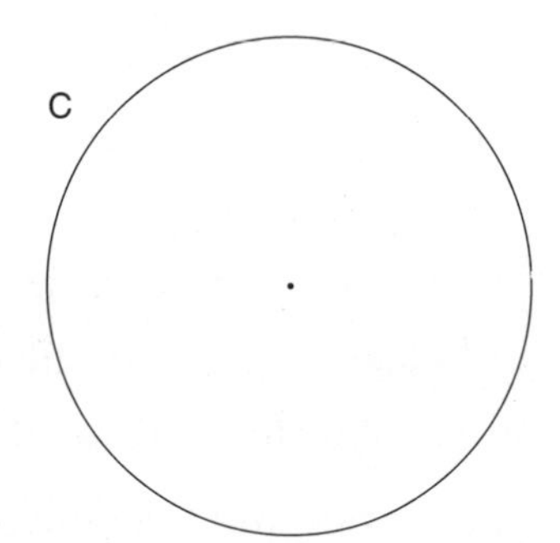

2 On centimetre square paper, draw a triangle with an area of $12\,\text{cm}^2$.

3 About 3000 years ago the Olmec people built a pyramid with a circular base.
The diameter of the base was 420 feet.
(a) How far was it to walk right round the base?
(b) What is this distance to the nearest 100 feet?

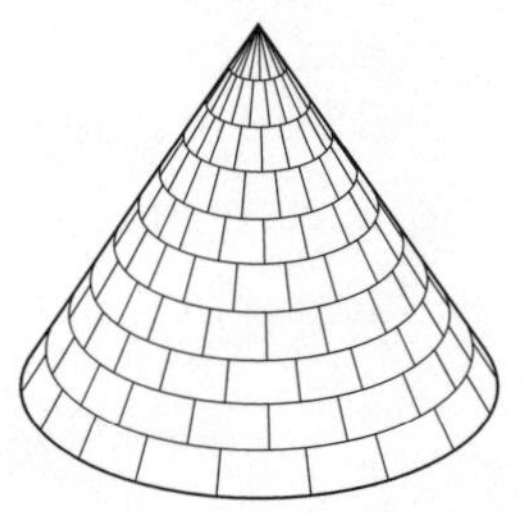

MEG/ULEAC (SMP)

Rounding ► page 4

4 An arrow for a traffic diversion sign is made using an isosceles triangle and a rectangle.

Calculate

(a) the perimeter of the arrow,

(b) the area of the triangle,

(c) the total area of the arrow.

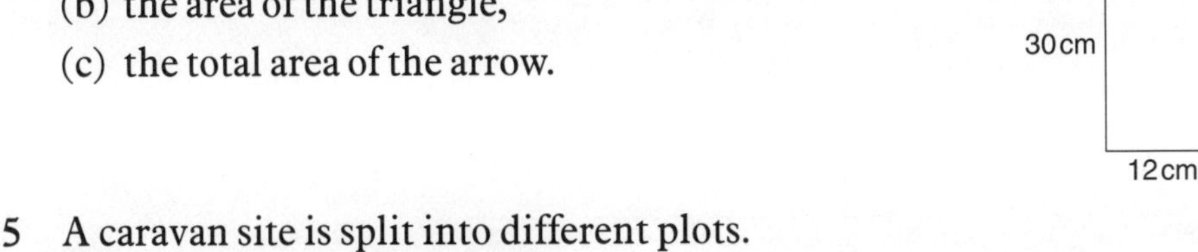

NICCEA

5 A caravan site is split into different plots.
These are two of the plots.

Calculate the area of each plot.

(a)

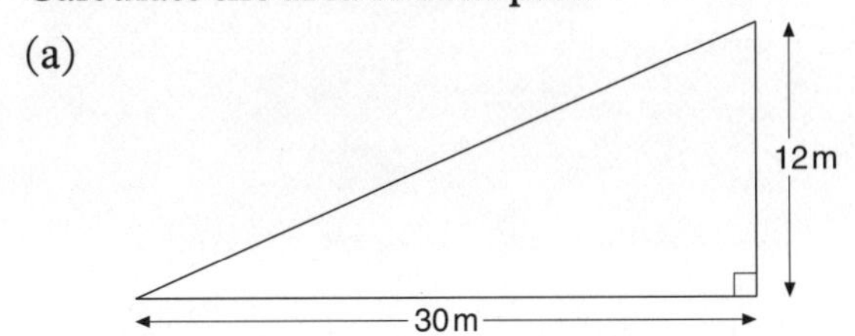

(b)

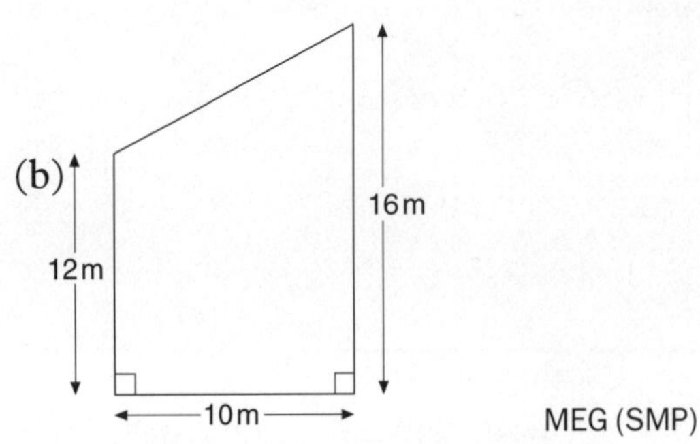

MEG (SMP)

6 In 1997 the Millennium wheel in London was planned to have a diameter of 151 metres.
80 cars were to be spaced equally around the circumference.

About how far apart would the cars be?

7 Sean has a tin of paint that will cover $20\,m^2$.
What is the radius of the largest circle he could cover in paint?

8 Bella wishes to make a circular pen for her pet rabbits.
She has 15 metres of wire netting.
What is the diameter of the largest pen she could make?

9 Sue makes a cake using a 10 inch diameter circular tin.

(a) The top of the cake is covered with a layer of marzipan.
What is the area of the top of the cake?

(b) A strip of marzipan 3 inches wide is put all around the edge of the cake.

(i) How long must the strip be?

(ii) What is the area of the strip?

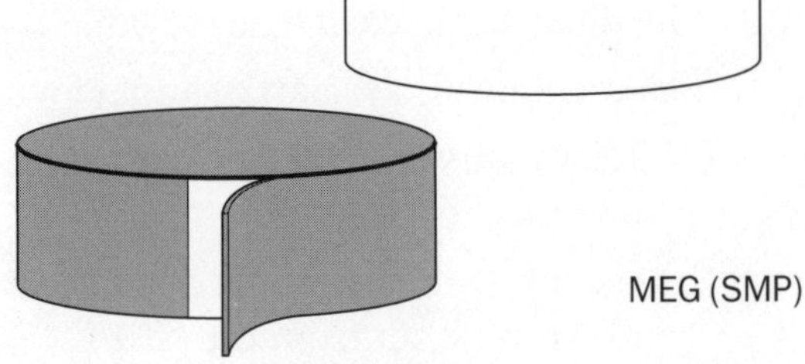

MEG (SMP)

Answers and hints ► page 129

Mixed shape, space and measures

1

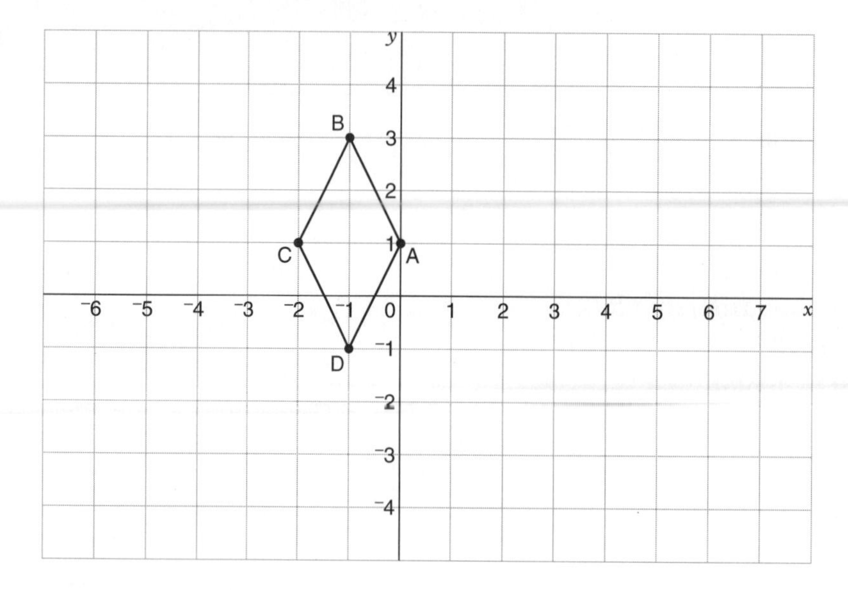

(a) Write down the coordinates of the points A, B, C and D.

(b) What special type of quadrilateral is ABCD?

(c) Draw at least six more of these quadrilaterals on a copy of the diagram, to form a tessellation.

MEG

2 These six squares form the net of a solid.

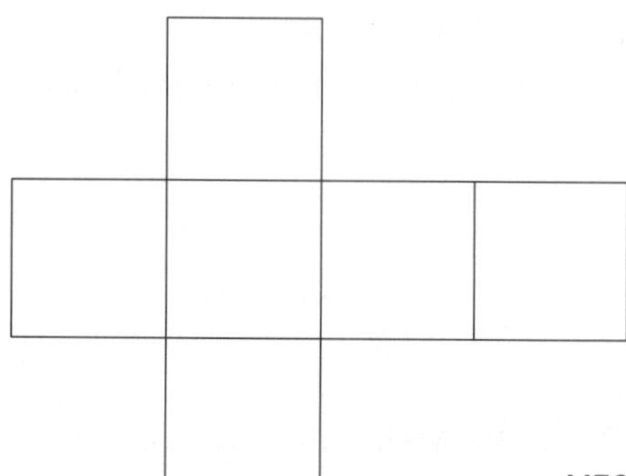

(a) Name the solid.

(b) Measure and write down the length of one side of a square in the net. (Your answer should be a whole number of centimetres.)

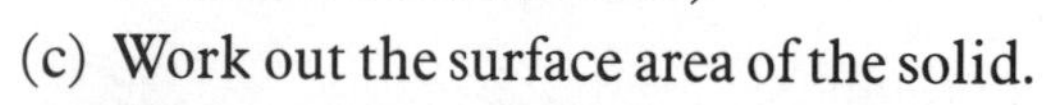

(c) Work out the surface area of the solid.

(d) Work out the volume of the solid.

MEG

3 A factory produces plastic name tags by stamping discs from rectangular sheets of plastic.

Each sheet of plastic is 40 cm long and 30 cm wide.

The diameter of each disc is 5 cm.

The diagram shows part of a sheet with discs cut from it.

(a) How many discs can be cut in this way from one sheet of plastic?

(b) Calculate the total area of all the discs that can be cut from the sheet.

(c) Calculate the area of plastic left over.

WJEC

4 Show any measurements or formulas you use in your working.
This tangram is made from seven pieces. It is drawn accurately.
Each piece is labelled with a letter.

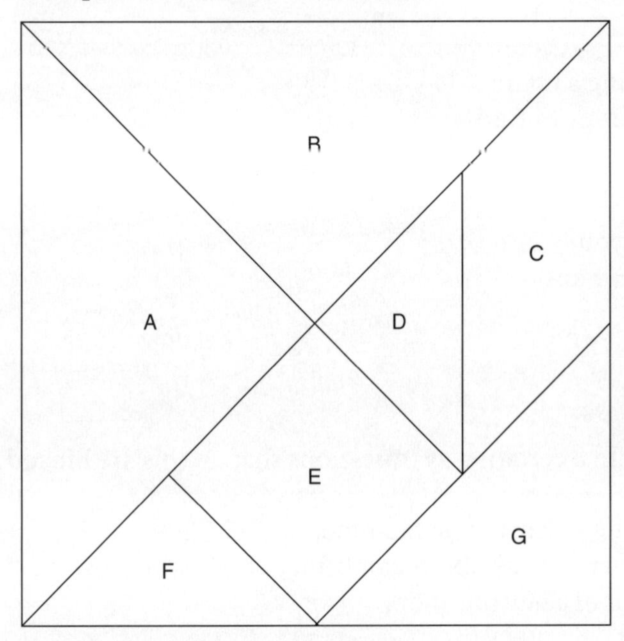

(a) All the 7 pieces together make a square.
Find the area of the square.
(b) Find the area of the triangular piece B.
(c) Find the area of the triangular piece D.
(d) Which piece is congruent to piece D?

MEG/ULEAC (SMP)

5 Meena has divided a rectangular sheet of paper into three triangles.

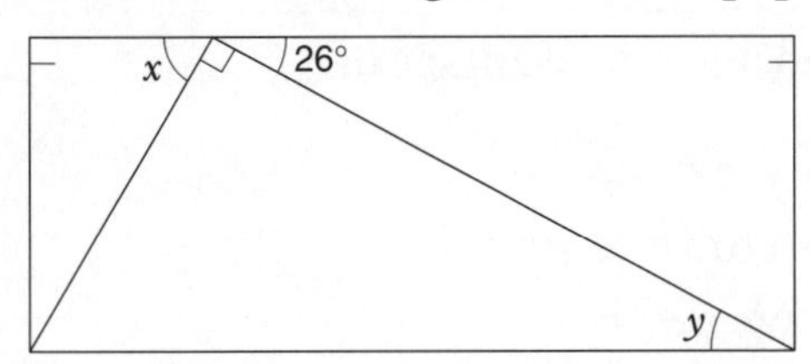

Not to scale

(a) Work out the size of angles x and y. Give reasons for your answers.

(b) Work out the area of the shaded triangle.

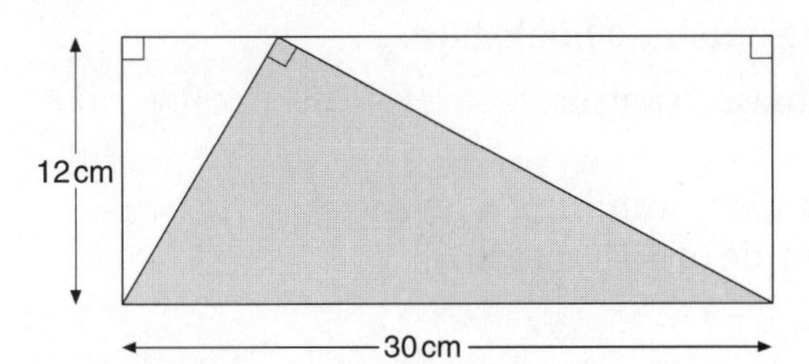

Not to scale

MEG (SMP)

Answers and hints ► page 130

HANDLING DATA

Collecting data

Canteen questionnaire

Name ______

1. How old are you?

11–13 ☐ 14–16 ☐ 17–18 ☐

2. When did you

Car colour observation sheet

Colour	Tally	Frequency
Red	⁄⁄⁄⁄ ⁄⁄⁄⁄ II	12
Black	II	2
Blue	III	

Data is usually collected for a purpose.
For example, a catering company planning a menu for a school canteen might carry out a survey to find out what young people like to eat at lunchtime.

Data is collected on **data collection sheets** like these.

For a questionnaire, you need to design your questions very carefully, especially if you want to show your results in a table or chart.
They should:

- be clear
- be easy to answer
- not try to encourage people to answer in a certain way (questions that do this are **biased**).

You also need to think carefully about who you ask. For example, if you want to find out what adults in general think about gambling, it's no good just asking adults coming out of a betting shop.
That would be a **biased sample** of adults.

1 Ken and Amy design questionnaires about school uniform.

Ken suggests: Which colour would you prefer for our sweatshirt? ______

Amy suggests: Which colour would you prefer for our sweatshirt?
Blue ☐ Green ☐ Red ☐ White ☐

For each question, give one advantage and one disadvantage of the way it is designed.

2 Jane asks this questions in a survey about a town centre:
'What do you think of the shops in the town centre?'
What is wrong with Jane's question?

3 Ameet wants to find out about people who use the local library.
He does his survey one Monday morning outside a bookshop.
Give two reasons why this would give a biased sample.

4 You are to carry out a survey on the time spent watching television last Sunday by collecting data from a sample of girls and boys of different ages.
Design an observation sheet to help you do this.

MEG

Answers and hints ► page 131

Timetables and calendars

Times can be written:
- in 12 hour time using a.m. or p.m.
- in 24 hour time

For example: 3:05 p.m. is 15:05 in 24 hour time
3:05 a.m. is 03:05 in 24 hour time

a.m. for a time before 12 noon but after 12 midnight
p.m. for a time after 12 noon but before 12 midnight

60 seconds = 1 minute
60 minutes = 1 hour
24 hours = 1 day
365 days = 1 year
(366 days = 1 leap year)

For days of the month:
30 days hath September, April, June and November,
All the rest have 31,
Excepting February alone,
Which has 28 days clear,
But 29 in each leap year.

1 Write these times using a.m. or p.m.

(a) 17:30 (b) 06:35 (c) 21:15 (d) 00:20

2 The times of some TV programmes on a Tuesday in 1997 are shown.

(a) How many minutes was *The Simpsons* on for?

(b) Sheila arrived home at 19:45. How long did she have to wait until the start of *Children's Hospital*?

(c) Gary recorded *Heartbreak High* and *EastEnders*. How much time did he use altogether on his videotape?

1	BBC1
5.35pm	Neighbours
6.00pm	News
6.30pm	News West
7.00pm	Holiday
7.30pm	EastEnders
8.00pm	Children's Hospital
8.30pm	999 Lifesavers

2	BBC2
5.50pm	Lifeline
6.00pm	The Simpsons
6.25pm	Heartbreak High
7.10pm	The O Zone
7.30pm	From the Edge
8.00pm	The House Detectives
8.30pm	Food

3 If 1 July is a Wednesday:

(a) Which day of the week is 10 July?

(b) Which day of the week is the last day in July?

4 This is a part of a timetable for trains from Carlisle to Preston.

Carlisle	1135	1204	1316	1355
Penrith	1152	1220	—	1411
Oxenholme	1218	1245	1353	1438
Lancaster	1235	1300	1409	1454
Preston	1259	1328	1432	1528

(a) Jean travelled from Carlisle to Preston on the 1355 train. How long did her train journey take?

(b) Helen arrived at Lancaster station at 2:00 p.m. and caught the next train to Preston. When did she arrive at Preston?

(c) Jamie needs to be in Preston by 2:15 p.m. What is the latest train he can catch from Penrith?

Answers and hints ► page 131

Interpreting tables

Information is often presented in tables to make it easier to understand.
You may be asked to find a piece of information, work out totals or compare quantities.

1 The information below is part of a computer database.

Name	*Age*	*Sex*	*Hair colour*	*Wears glasses*
Anne	23	F	Fair	No
Bill	25	M	Brown	No
David	19	M	Brown	Yes
Gary	22	M	Black	No
Mary	22	F	Brown	No
Chris	18	F	Fair	Yes
Rashid	24	M	Brown	No
Sue	19	F	Red	No

(a) What colour is Sue's hair?

(b) How many people in this group are female?

(c) List all the people under 21.

(d) List all the males with brown hair.

(e) What fraction of this group of people wear glasses? MEG

2 This table shows the cost of posting letters in 1997.

(a) How much was it to send a 50g letter first class?

(b) What was the cost of sending a 220g letter second class?

Weight *up to*	First class	Second class
60g	26p	20p
100g	39p	31p
150g	49p	38p
200g	60p	45p
250g	70p	55p
300g	80p	64p
350g		

3 The mileage chart shows the distances in miles between four towns.

(a) How far is it from Exeter to London?

(b) Sarah drives from Glasgow to London. Tim drives from Glasgow to Exeter. Who travels further?

Exeter			
456	Glasgow		
200	414	London	
267	189	225	Preston

4 The two-way table shows information about the number of students studying 'A' level English and 'A' level Mathematics in a school.

(a) How many girls study 'A' level Mathematics?

(b) How many students study 'A' level English?

	Boys	Girls
Mathematics	20	10
English	28	60

Answers and hints ► page 132

Graphs and charts

A **pictogram** can be an attractive way to display information.
You need to know what each symbol stands for.

For example, if ◯ stands for 20 cakes, then ◖ stands for 10 cakes and for 15 cakes.

1 The pictogram shows the number of burgers sold by four students at a fete.

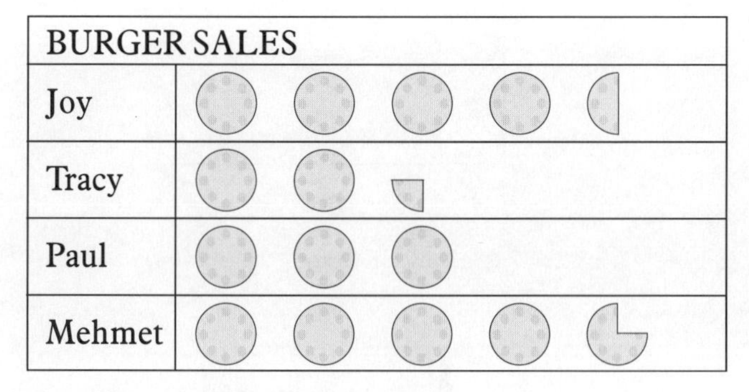

◯ represents 8 burgers

(a) Who sold the most burgers?

(b) How many burgers did Tracy sell?

(c) How many burgers did the four students sell in total?

Question 2 is on worksheet F18.

Tables or **bar charts** can be used to display information.
Some school students collect information about flavours of crisps sold at break.

Flavour	Tally	Frequency
Plain	卌 卌 ||	12
Vinegar	卌 卌	10
Cheese	卌 卌 卌 ||||	19
Bacon	卌 |||	8
Beef	卌 |	6

Frequency here means the number of bags of crisps.

In a tally chart 卌 stands for 5 items.

Each of these bar charts shows the information.

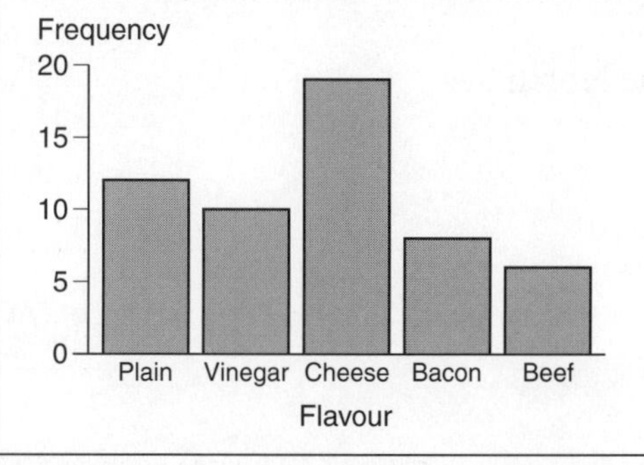

Sometimes, this sort of chart is called a **stick graph**.

Frequency
20
15
10
5
0
Plain Vinegar Cheese Bacon Beef
Flavour

3 A box contains sweets of different colours.

The bar chart shows how many sweets of each colour are in the box.

(a) How many red sweets are in the box?

(b) How many blue sweets are in the box?

(c) How many sweets are in the box altogether?

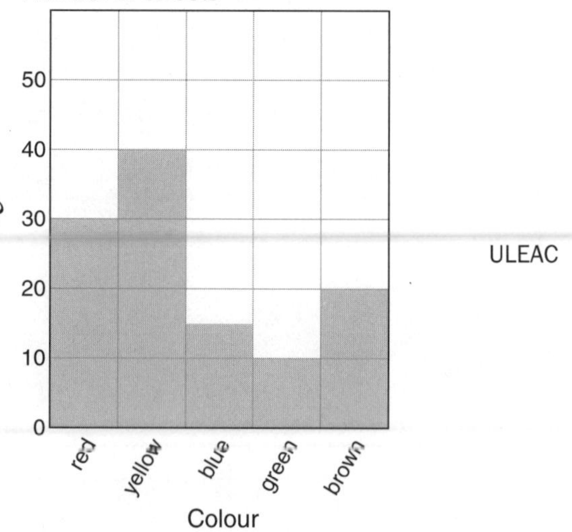

ULEAC

4

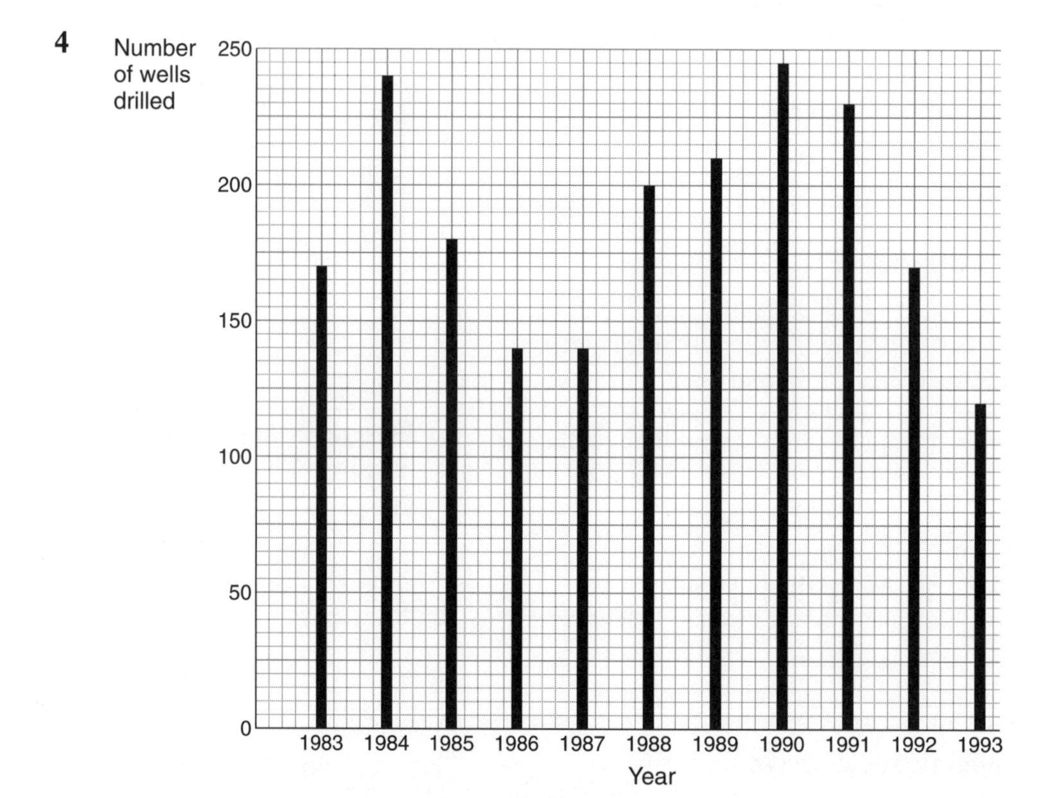

The chart shows the number of oil wells drilled in the North Sea between 1983 and 1993.

(a) In which year were most wells drilled?

(b) How many wells were drilled in 1988?

(c) How many wells were drilled in 1991?

ULEAC

5 The bar chart shows the shoe sizes of a group of 12-year-old girls.

(a) How many girls have a shoe size larger than 3?

(b) Comment on the shape of the bar chart, saying whether or not this is the shape you would expect.

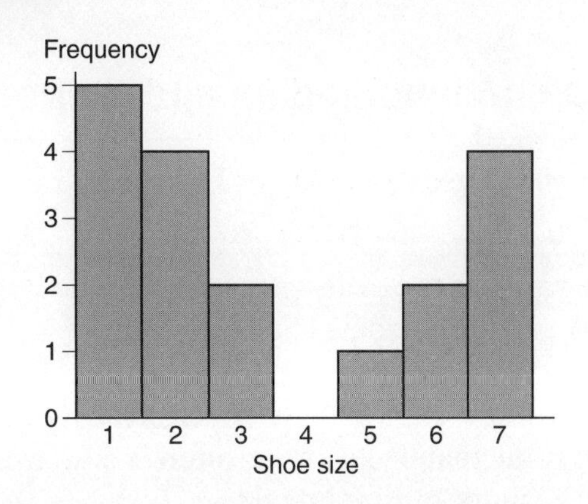

Questions 6, 7 and 8 are on worksheets F18 to F20.

It is best to show some types of information as a **line graph.** **Drawing graphs ► page 31**

The table shows Jan's height measured at various ages between 1 and 5.
A line graph is drawn for these results.

Age (years)	Height (cm)
1	74
2	86
3	95
4	102
5	109

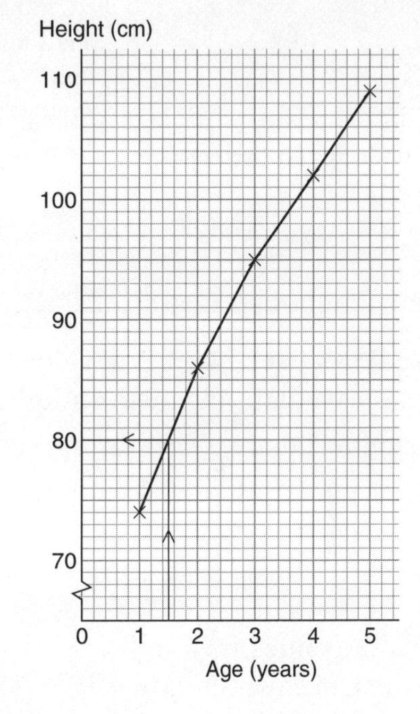

The graph can be used to estimate heights.
For example:
Jan's height at age $1\frac{1}{2}$ years was about 80 cm.

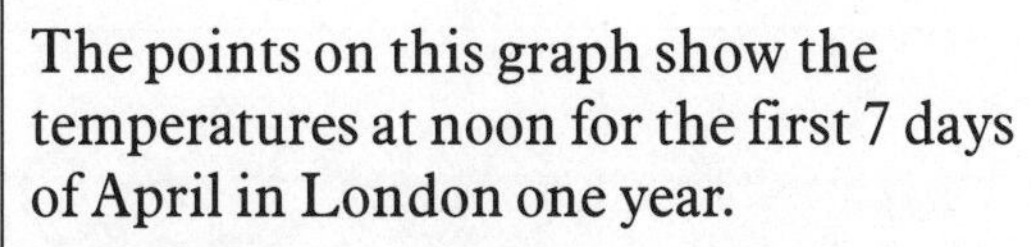

The points on this graph show the temperatures at noon for the first 7 days of April in London one year.

The points are joined up to show the shape of the graph.
The line is dotted to show that you could not use the graph to estimate, for example, the temperature at midnight on 1 April.

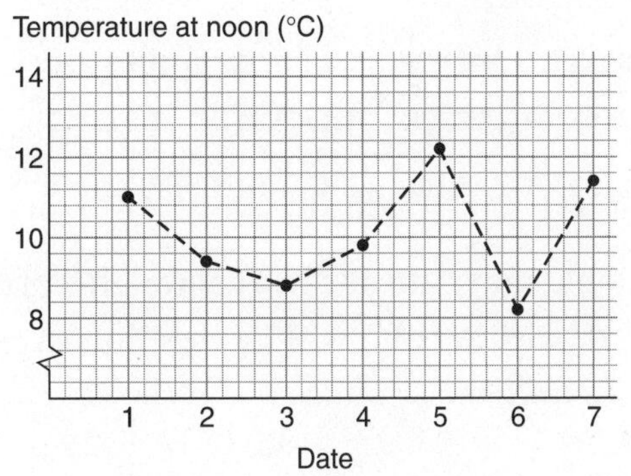

Question 9 is on worksheet F21.

Answers and hints ► page 132

Mode, mean, median and range

The shoe sizes of a group of 15 women are:

6 5 7 6 5 6 6 6 8 4 5 7 6 7 3

The **range** is the difference between the largest and smallest value.
So the range for these shoe sizes is $8 - 3 = 5$.

Range is a measure of spread.

The **mode** is the value that occurs most often.
The value that occurs most often is 6, so we say:
the mode of the shoe sizes is 6
or the **modal** shoe size is 6.

To find the **mean** shoe size, add up the values and divide by the number of values.
The total is: $6 + 5 + 7 + 6 + 5 + 6 + 6 + 6 + 8 + 4 + 5 + 7 + 6 + 7 + 3 = 87$
$87 \div 15 = 5{\cdot}8$, so the mean shoe size is 5·8.

To find the **median** shoe size, put the values in order and find the middle one.
Put the values in order:

3 4 5 5 5 6 6 6 6 6 6 7 7 7 8

middle value

The median shoe size is 6.

If there are an **even** number of values, there will be two values in the middle.
To find the median, add up the middle values and divide by 2.
For example, the shoe sizes for a group of 10 men are:

6 7 7 8 8 9 9 9 10 11

middle values

The middle values are 8 and 9,
so the median shoe size is $(8 + 9) \div 2 = 8{\cdot}5$.

Mode, mean and median are measures of average.

1 The weights, in grams, of nine apples are:

100 123 99 107 102 118 121 106 103

What is the median weight of these apples?

2 This table shows the individual weights of 10 Victoria plums and 10 Merryweather plums. The weights are given in grams.

Victoria	23	36	37	29	24	40	31	20	27	33
Merryweather	23	26	27	26	24	27	24	27	21	25

Calculate the mean weight of each type of plum. WJEC

3 (a) The weights, in kilograms, of the 8 members of Hereward House tug of war team at a school sports are:

75 77 77 76 84 76 77 78

Calculate the mean weight of the team.

(b) The 8 members of Nelson House tug of war team have a mean weight of 64 kilograms.

Which team do you think will probably win a tug of war between Hereward House and Nelson House?

Give a reason for your answer: MEG

4 The weights, in kilograms, of members of a rowing crew are

80, 83, 83, 86, 89, 91, 93, 99.

(a) Calculate the mean of these weights.

(b) Calculate the range of these weights. MEG

5 A shop assistant sold walking boots to 10 customers.
The sizes of the boots were:

5 6 6 7 7 8 8 8 8 10

Which is the modal boot size?

6 The heights (in cm) of ten 1-year-old children are:

69 73 65 78 70 68 72 74 67 74

(a) Calculate the mean height.

(b) What is the median height?

(c) (i) Which height is the mode?

(ii) Why is the mode not a good measure of the average for these heights?

7 There are 24 people in an old people's home.
There are 15 women.
Their ages are shown below.

98 85 87 75 71 69 76 82
93 94 78 77 91 90 79

(a) (i) What is the range of the ages of the women?

(ii) Calculate the mean age of the women.

All the people in the home are 65 years old or more.
The men have a mean age of 71 years.
The range of the ages of the men is 34 years.

(b) What is the sex of the oldest person in the home?
Give a reason for your answer. SEG

8 (a) The weekly wages of 10 workers, in pounds, at the Bentgate factory are:

170, 147, 170, 138, 149, 163, 162, 162, 178, 561.

(i) Calculate the median wage.

(ii) Calculate the mean wage.

(b) The weekly wages of the 11 workers, in pounds, at the Penlight factory are:

177, 220, 159, 200, 143, 150, 190, 172, 175, 189, 227.

(i) Calculate the median wage.

(ii) Calculate the mean wage.

(c) Which is the better average to compare the wages in the two factories, the median or mean?
Give a reason for your answer.

MEG (SMP)

9 Leslie and Pat are members of a quiz team.
Here are their scores for the last eight quizzes:

Leslie	79	73	75	91	74	84	76	80
Pat	78	75	79	86	76	81	77	80

(a) Calculate Leslie's mean score.

(b) Find the range of Leslie's scores.

Pat has a mean score of 79 and a range of 11.

(c) If you were captain of the quiz team, which of these players would you choose?
Give a reason for your choice.

MEG/ULEAC (SMP)

10 Eight judges each give a mark out of 6 in an ice-skating competition.
Oskana is given the following marks:

5·3 5·7 5·9 5·4 4·5 5·7 5·8 5·7

The mean of these marks is 5·5, and the range is 1·4.
The rules say that the highest and lowest marks are to be deleted:

5·3 5·7 ~~5·9~~ 5·4 ~~4·5~~ 5·7 5·8 5·7

(a) (i) Find the mean of the six remaining marks.

(ii) Find the range of the six remaining marks.

(b) Do you think it is fairer to count all eight marks, or to count only the six remaining marks?

Use the means and ranges to explain your answer.

(c) The eight marks obtained by Tonya in the same competition have a mean of 5·2 and a range of 0·6.

Explain why none of her marks could be as high as 5·9.

MEG

Answers and hints ► page 134

Grouped data

Tom carries out a survey on lateness at his school.
He finds the number of lates for each student last term.

Here are the results for class 9T.

0	3	7	2	11	8	5	25	4	0
0	0	6	1	2	0	5	12	1	3
7	5	0	2	16	0	0	4	2	

This data can be organised in a frequency table using groups.
For example:

Number of lates	Tally	Frequency
0–4	𝍸 𝍸 𝍸 |||	18
5–9	𝍸 ||	7
10–14	||	2
15–19	|	1
20–24		0
25–29	|	1

1 Sharon conducted a survey at her Youth Club.
She asked each member how many hours they had spent watching television last week.

Here are her results.

4	27	14	22	12	24	10	26
25	19	7	29	17	23	16	18
11	17	25	19	15	8	25	20
23	14	9	20	13	30	18	

(a) Copy and complete this frequency table for these results.

Number of hours watching television	Tally	Frequency
1–5		
6–10		
11–15		
16–20		
21–25		
26–30		

(b) (i) How many members were surveyed?

(ii) How many members had watched more than 15 hours of television last week?

MEG/ULEAC (SMP)

Answers and hints ► page 135

Grouped data and frequency diagrams

Some students were asked how many hours they spent doing homework during one week.
The results are shown in the frequency table below.

This means 5 hours or more but less than 10 hours.

Time (hours)	Frequency
$0 \leq$ time < 5	3
$5 \leq$ time < 10	22
$10 \leq$ time < 15	35
$15 \leq$ time < 20	11

A **frequency diagram** for the data can be drawn.

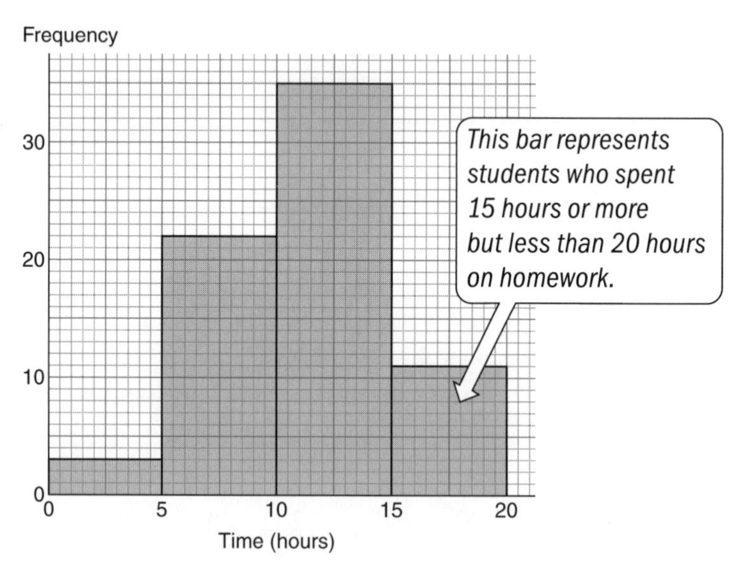

A **frequency polygon** can be drawn by joining the mid-points on the tops of the bars.

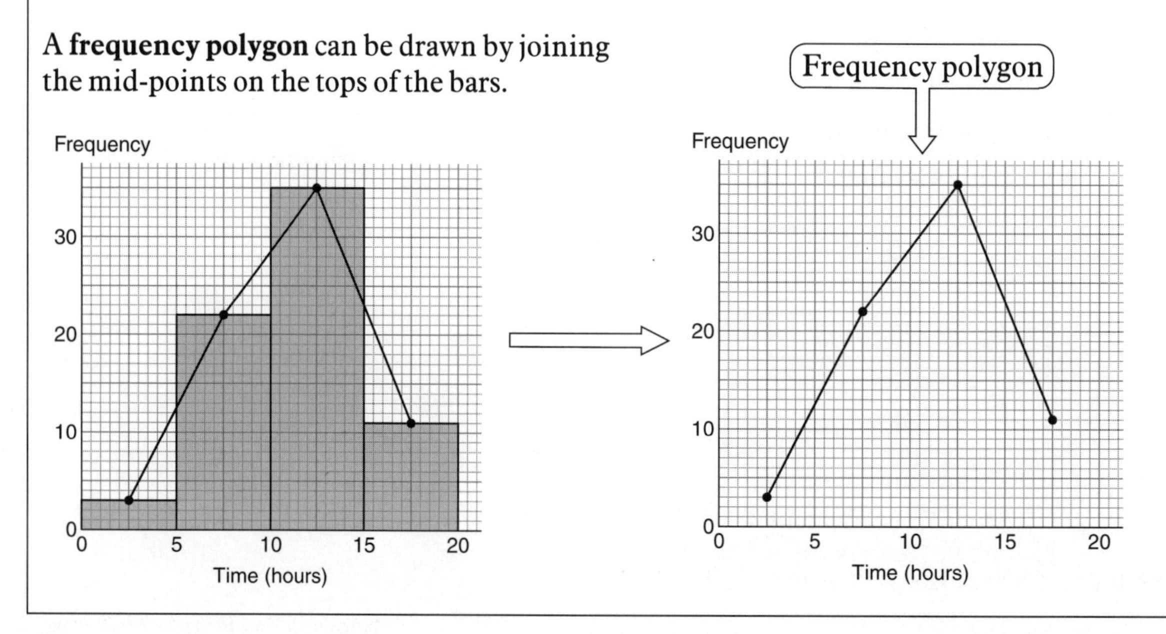

1 Zara measured the heights of 30 pupils in her class.
She measured the heights in centimetres.
Here are her results.

155 147 156 172 152 159 165 154 163 171
166 144 172 168 160 162 182 164 167 173
164 168 154 163 173 149 157 153 162 170

(a) She realised it would be easier to group her results.
Copy and complete the table below.

Height (centimetres)	Tally marks	Number of pupils
140 and less than 150		
150 and less than 160		
160 and less than 170		
170 and less than 180		
180 and less than 190		

(b) How many pupils were 170 cm or taller?

MEG (SMP)

2 Jane collected the heights of some women in metres.
The heights are

1·49 1·62 1·58 1·68 1·71 1·56 1·65 1·63 1·75 1·71
1·53 1·55 1·72 1·68 1·49 1·63 1·62 1·67 1·70 1·60
1·64 1·68 1·79 1·63 1·60 1·66 1·67 1·69 1·62 1·68

An observation sheet has been started using equal class intervals for the heights.

(a) Copy and complete the observation sheet.

Height (metres)	Tally	Frequency
$1{\cdot}40 \leq$ height $< 1{\cdot}45$		
$1{\cdot}45 \leq$ height $< 1{\cdot}50$		
$1{\cdot}50 \leq$ height $< 1{\cdot}55$		
$1{\cdot}55 \leq$ height $< 1{\cdot}60$		
$1{\cdot}60 \leq$ height $< 1{\cdot}65$		
$1{\cdot}65 \leq$ height $< 1{\cdot}70$		

(b) How many women were shorter than 1·60 m?

Questions 3 to 5 are on worksheets F22 to F25.

Answers and hints ► page 135

Pie charts (using a pie chart scale)

Drawing pie charts

Each small division on a pie chart is 1%.

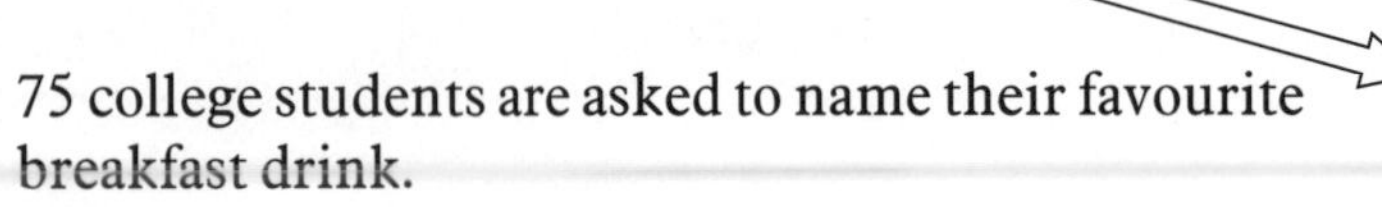

75 college students are asked to name their favourite breakfast drink.
The results are shown in the table.

The percentages can be calculated for the results.

Percentages ► page 18

Drink	Number of students
Tea	42
Coffee	21
Milk	3
Juice	9
Total	75

Percentages

$\frac{42}{75} \times 100 = 56\%$

$\frac{21}{75} \times 100 = 28\%$

$\frac{3}{75} \times 100 = 4\%$

$\frac{9}{75} \times 100 = 12\%$

Check that your percentages give a total of 100%.

A pie chart scale can now be used to draw a pie chart.

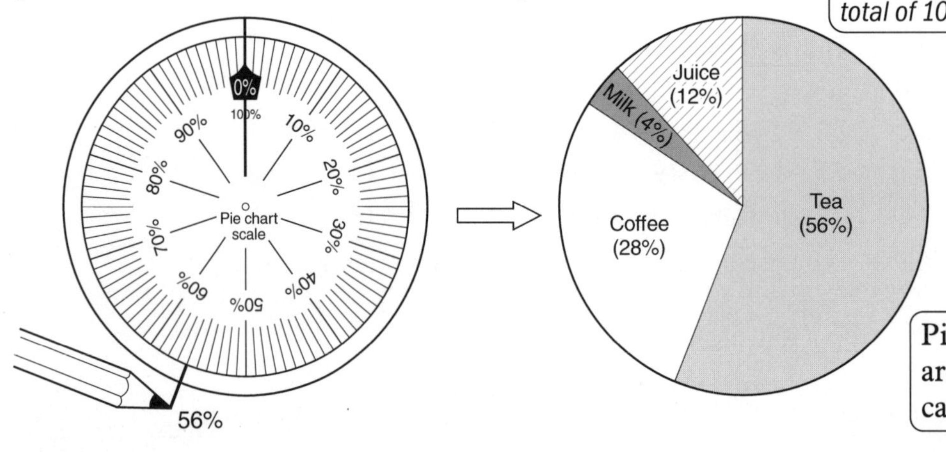

Pie chart 'slices' are sometimes called **sectors**.

Interpreting pie charts

180 cars were sold at an auction.
The pie chart shows the countries where they were made.

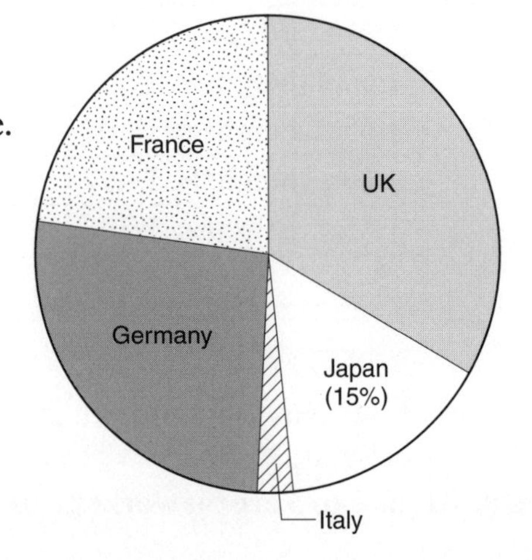

Using a pie chart scale, the percentage of cars made in Japan is found to be 15%.

So the number of cars made in Japan is
15% of 180 cars = $0{\cdot}15 \times 180$ cars = 27 cars.

The fraction of cars made in Japan is $\frac{15}{100} = \frac{3}{20}$.

Fractions ► page 14

1 In 1993 the National Trust had a total income of £110 million.

Source of income	Amount in millions	%
Membership	40·7	37
Inheritance	19·8	
Investment		
Rents	13·2	
Other	17·6	16
Total	110·0	100

(a) Copy and complete the table.

(b) Draw a pie chart to show how the National Trust raised its income in 1993. (Use worksheet F26.)

MEG (SMP)

2 A survey was carried out to find where the members of Parkside Middle School spent their holidays last summer.

The results for the boys are shown in the table.

Place	Boys	%
Britain	110	
France	75	
Spain	15	
Rest of Europe	40	
Other	10	
Total	250	

(a) Copy and complete the table.

(b) Draw a pie chart to show where the boys at the school spent their holidays. (Use worksheet F26.)

MEG (SMP)

3 3800 people were killed in road accidents in1993.
The pie chart shows who they were.

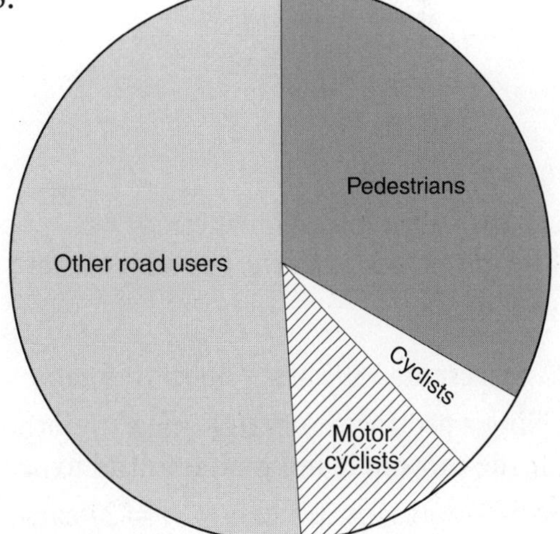

(a) What percentage of the people who died were motor cyclists?

(b) How many pedestrians died in road accidents in 1993?

(c) What fraction of the people who died were cyclists?

Answers and hints ► page 137

Pie charts (using a protractor)

Drawing pie charts

90 college students are asked to name their favourite sandwich.
The results are shown in the table.

The angles can be calculated for the results.

360° is divided between the 90 students to give $360^\circ \div 90 = \mathbf{4^\circ}$ for each student.

Sandwich	Number of students
Cheese	10
Chicken	15
Beef	27
Ham	38
Total	90

Angles

$4^\circ \times 10 = 40^\circ$

$4^\circ \times 15 = 60^\circ$

$4^\circ \times 27 = 108^\circ$

$4^\circ \times 38 = 152^\circ$

Check that your angles give a total of 360°.

A protractor or angle measurer can now be used to draw a pie chart.

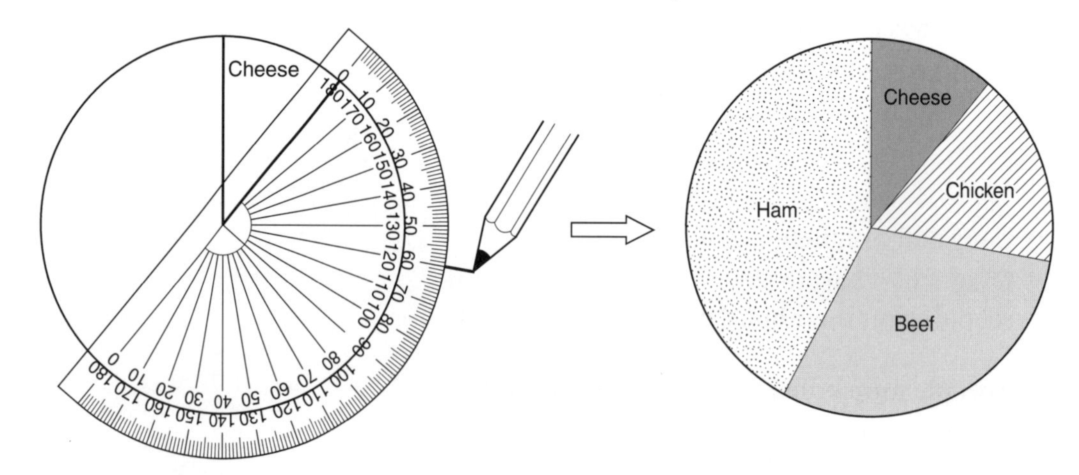

Interpreting pie charts

180 cars were sold at an auction.
The pie chart shows the countries where they were made.

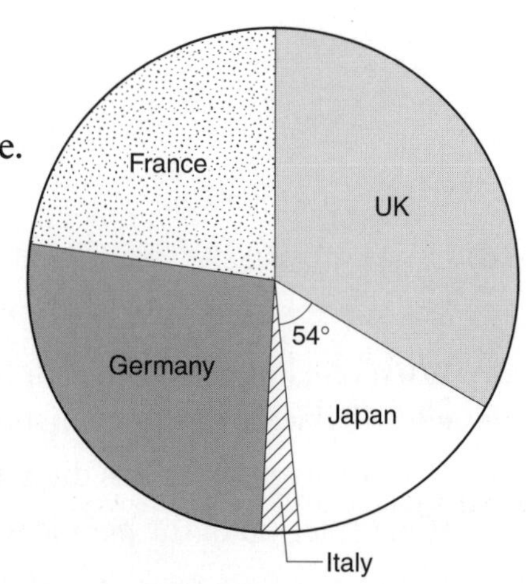

1° represents 180 cars $\div$ 360 = 0·5 cars.

Using a protractor or angle measurer, the angle for the cars made in Japan is found to be 54°.

So 54° represents 0·5 cars $\times$ 54 = 27 cars.

The fraction of cars made in Japan is $\frac{54}{360} = \frac{3}{20}$.

Fractions ► page 14

1 A car dealer sold 120 new cars in August.
She made a note of the colour of the cars she sold.

(a) Copy and complete the table below to show the pie chart angles for each colour.

Car colour	Number of cars	Pie chart angle
Red	55	
Blue	32	
White	20	
Grey	13	
Total	120	

(b) Show this information on a pie chart. (Use worksheet F26.) MEG

2 The pie chart shows how Polly spends one whole day.

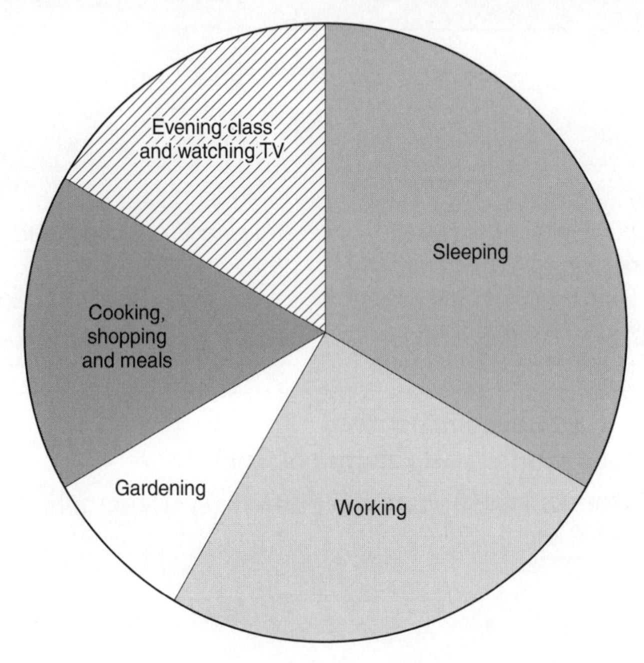

(a) What fraction of the day did Polly spend working?

(b) How long did she spend gardening?

(c) She spent 1 hour and 30 minutes watching TV.
How long did she spend at her evening class?

(d) How many hours did she sleep for?

Questions 3, 4 and 5 are on worksheets F27 and F28.

Answers and hints ► page 138

Scatter diagrams

A **scatter diagram** (or scatter graph) is used to see if there appears to be a link between two features: for example, the heights and weights for a group of people.

Pairs of values are plotted as points.
If this gives a pattern where points lie close to a line we say there is **correlation** between the features.

Here are three examples that show different patterns of correlation.

Positive correlation

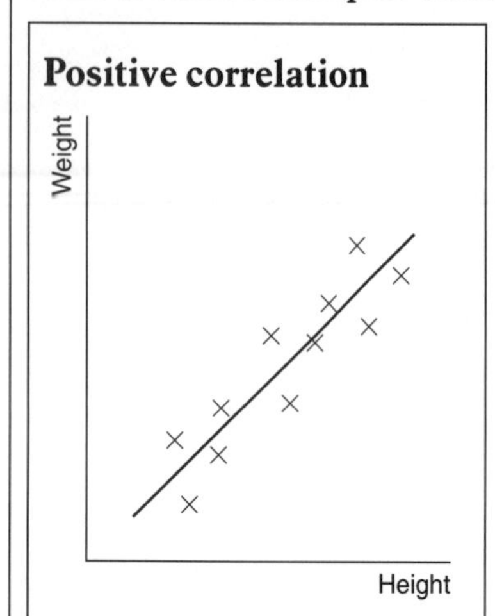

Points appear to lie close to a line sloping up to the right. As height goes up, weight goes up.

Negative correlation

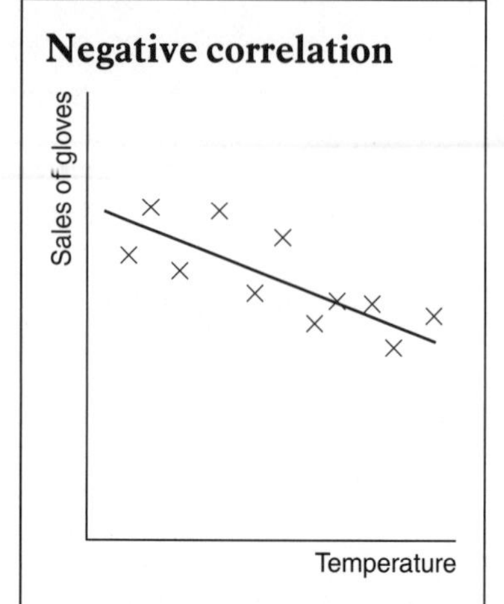

Points appear to lie close to a line sloping down to the right. As temperature goes up, sales of gloves generally go down.

No correlation

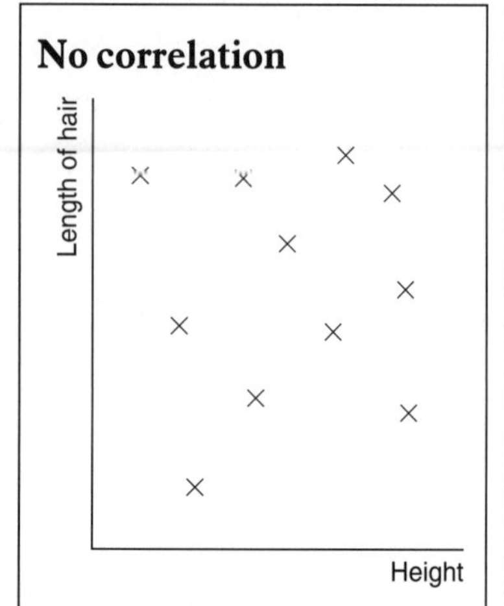

No clear line is suggested from the points. There seems no link between height and length of hair.

1 Megan wanted to find out if there is a connection between the average temperature and the total rainfall in the month of August.

She obtained the weather records for the last 10 years and plotted a scatter graph.

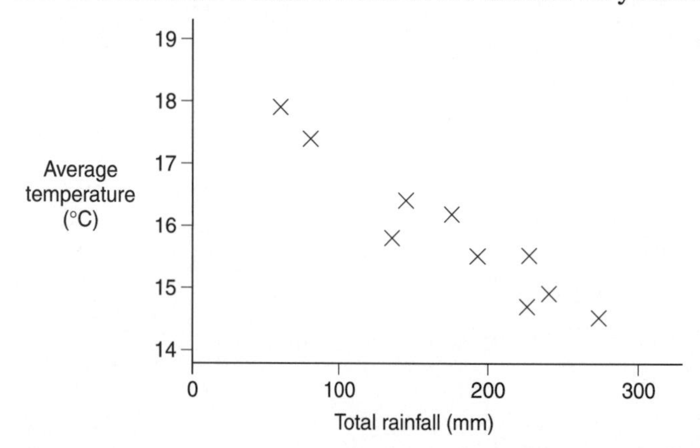

What does the graph show about a possible link between temperature and rainfall in August?

MEG

2 Jason asked a group of pupils to say how much time they spent revising for a test and also the mark they scored out of 10.
Here is a scatter graph to show the results.

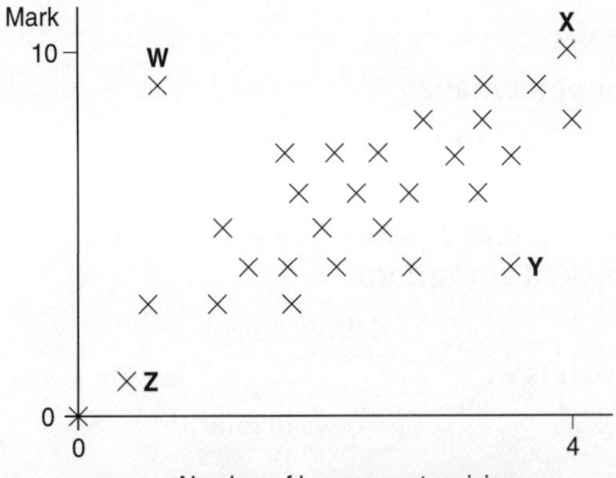

(a) Which of the four points W, X, Y or Z matches each of the students below?

Paul: 'I spent a whole evening revising and scored full marks.'

Spencer: 'I hardly did any revision and I did badly in the test.'

Liz: 'I did well in the test and only spent a short time revising.'

(b) What does the graph suggest about the relationship between the time spent revising and the mark scored?

3 Some students were drawing scatter diagrams.

- Sue drew a graph to show *a person's height* and *their score in a French test*.
- Aisha drew a graph to show the *amount of rain* and *ice cream sales*.

Which of the graphs below would you expect for:

(a) Sue's graph (b) Aisha's graph?

Give a brief reason for each choice.

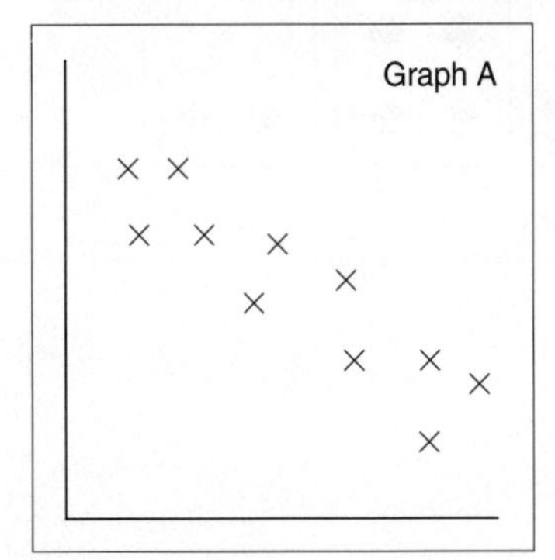

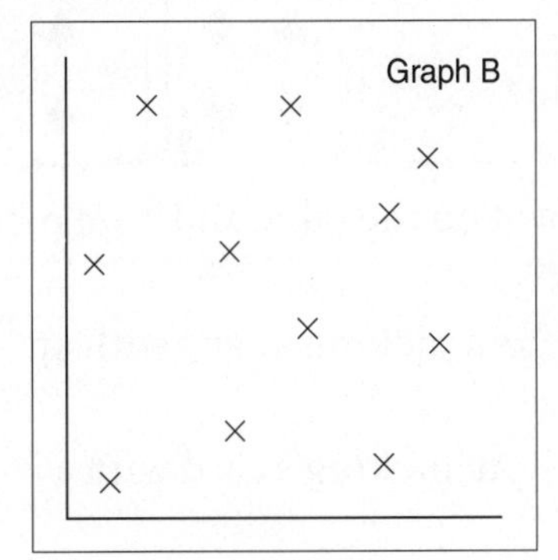

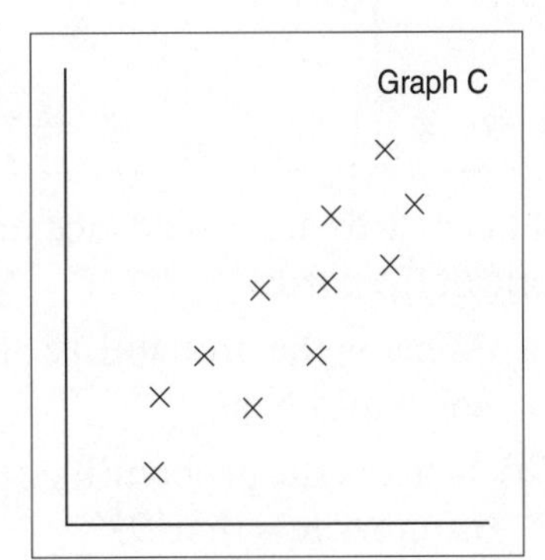

Questions 4 and 5 are on worksheets F29 and F30.

Answers and hints ► page 140

Probability

The **probability** of something happening is a value that tells you how likely it is to happen.

A probability of 0 means it can never happen.
A probability of 1 means it is certain to happen.

You can give a probability as a fraction, decimal or percentage.

Sometimes, you can work out probabilities without doing an experiment.

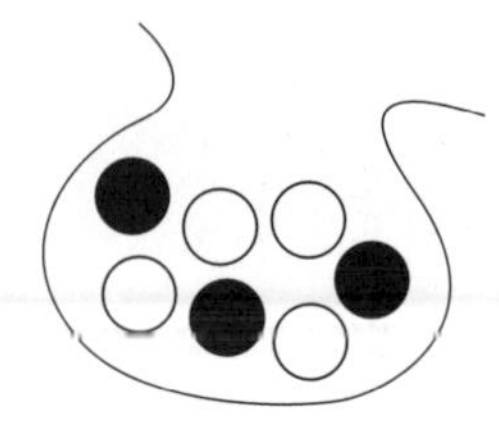

This bag contains 7 identically shaped balls so each ball has an equal chance of being picked at random.

So picking at random,
the probability of a white ball is $\frac{4}{7}$.

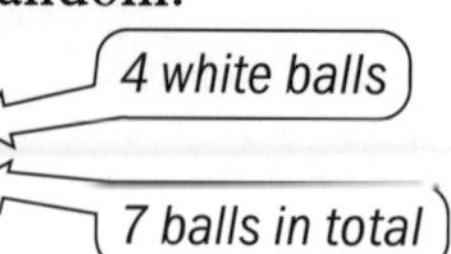

If you have done an experiment a lot of times you can **estimate** a probability from it.

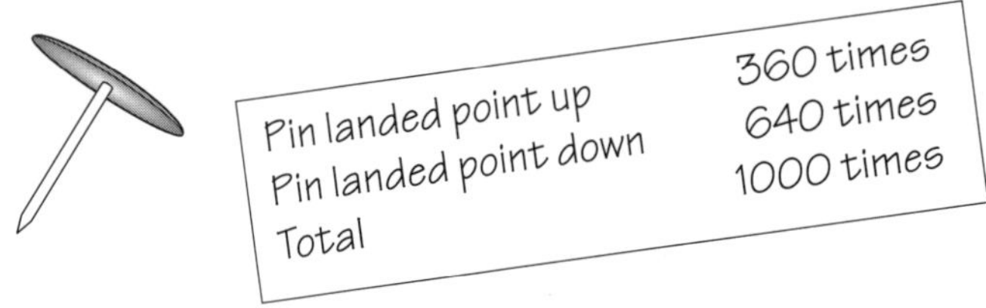

The probability of the pin landing point up can be estimated as $\frac{360}{1000} = 0{\cdot}36 = 36\%$.

This is sometimes called **experimental probability.**

The probability of something happening added to the probability of something not happening always gives 1 (or 100%).
For example, if the probability of it raining tomorrow is 25%, then the probability of it **not** raining is 75%.

Fractions, decimals and percentages ► page 16

1

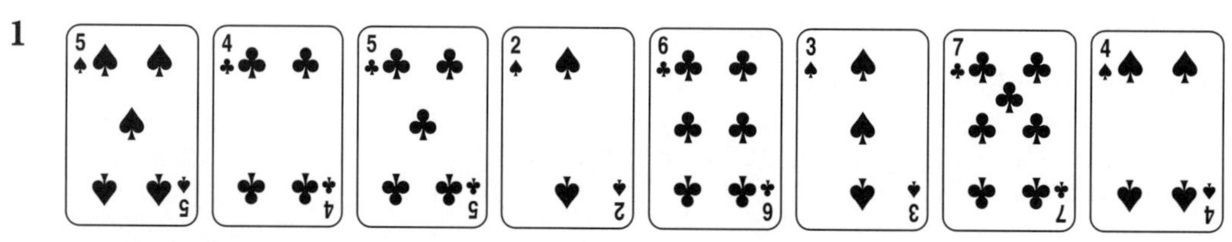

These cards are placed face down on the table and Tom picks one of the cards.

(a) What is the probability of Tom picking a card with an even number?

(b) What is the probability of Tom picking a card with a number less than 9?

Copy the probability scale.

0 ———————— 1

(c) Mark with a cross (×) the chance of Tom picking a card with a number bigger than 3. Explain your answer.

MEG/ULEAC (SMP)

2 A six-sided dice with the numbers 1, 2, 3, 4, 5 and 6 on its faces is rolled once.
What is the probability of rolling a number less than 3?

3 'The probability of winning a prize on the National Lottery with one ticket is $\frac{1}{50}$.'
From the statement above, what is probability, with one ticket, of **not** winning a prize on the National Lottery?

4 In a **Pick a straw** competition, there are 500 straws in a jar.
Each straw has a ticket rolled up inside it.

1 of the tickets wins a prize of £10.
4 of the tickets win prizes of £1.

Suppose you have first pick.

(a) What is the probability of winning the £10 prize?

(b) What is the probability of winning **any** prize? MEG/ULEAC (SMP)

5 A letter has a first-class stamp on it.
The probability that the letter will be delivered on the next working day is 0·86.
What is the probability that the letter will **not** be delivered on the next working day?

ULEAC

6 Girls and boys in Form 12TY have either dark hair or fair hair.
The table shows the numbers in each group.

Form 12TY	Dark hair	Fair hair
Girls	4	10
Boys	9	5

(a) How many pupils are in Form 12TY?

(b) What is the probability that a pupil chosen at random
  (i) is a girl,
  (ii) has fair hair,
  (iii) is a boy with fair hair? NICCEA

7 One of the stalls at a summer fair is a **Roll a penny** game.
Leo has recorded whether people win or lose when they play the game.

	Tally	Frequency
Win	𝍸 𝍸 I	
Lose	𝍸 𝍸 𝍸 𝍸 𝍸 𝍸 𝍸 𝍸 III	
	Total	

(a) (i) Copy and complete the frequency column in the table.
  (ii) What is the experimental probability that you win when playing this game?

(b) If you played this game 100 times, about how many times would you expect to win? MEG (SMP)

Answers and hints ► page 141

All the possible ways

Be methodical when listing all the different ways of arranging things.

This list shows all the possible ways of arranging the letters 'A', 'T' and 'E' in a row.

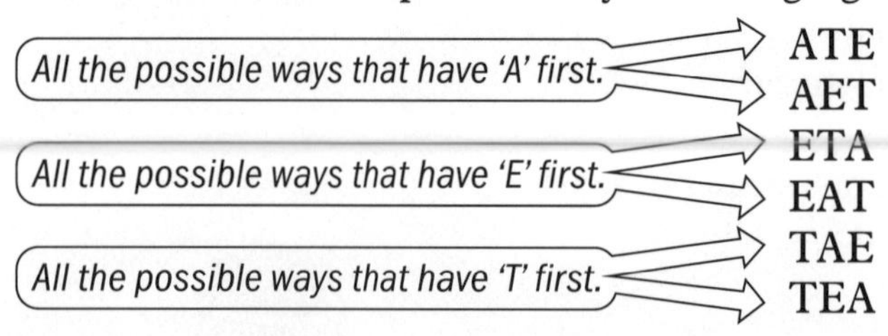

There are 6 different ways to arrange these letters.
If they are arranged at random, the probability of making the word 'TEA' is $\frac{1}{6}$.

If you spin these two fair spinners, there are many different possible results.

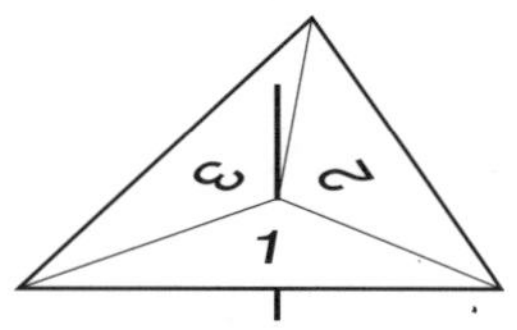

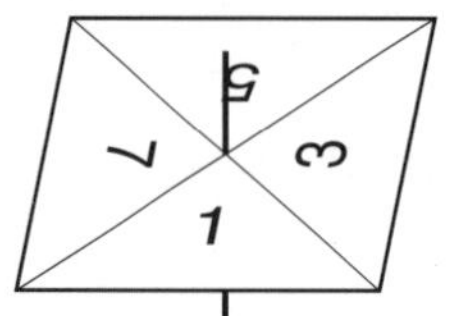

You can show the possible results in a table like this.
There are 12 different possible results.

Triangular spinner \ Square spinner	1	3	5	7
1	**1, 1**	**1, 3**	**1, 5**	**1, 7**
2	**2, 1**	**2, 3**	**2, 5**	**2, 7**
3	**3, 1**	**3, 3**	**3, 5**	**3, 7**

This shows a result of 3 on the triangular spinner and 5 on the square spinner.

To find the probability of a particular total score when both numbers are added, it helps to put the total scores in a table.

Triangular spinner \ Square spinner	1	3	5	7
1	**2**	**4**	**6**	**8**
2	**3**	**5**	**7**	**9**
3	**4**	**6**	**8**	**10**

The probability of a total score of 2 is $\frac{1}{12}$.

The probability of a total score of 6 or more is $\frac{7}{12}$.

1

Pat has these three cards.

Copy and complete the list to show all the different ways that she can arrange them.

P	A	T
P	T	A

2 Helen has three T-shirts: a white one, a red one and a green one.
She has two pairs of shorts: a blue pair and a white pair.

(a) Copy and complete this list to show all the different choices of T-shirt and shorts she has.

T-shirt	Shorts
White	Blue

(b) If she picks a T-shirt and pair of shorts at random, what is the probability that they will both be white?

3 In a restaurant there is a choice of:

- three main courses – Steak pie (S), Fish (F), Vegetable lasagne (V), and
- four puddings – Ice cream (I), Trifle (T), Apple pie (A), Cheesecake (C).

(a) Copy and complete the list to show all possible combinations of the two courses.

Main course	Pudding
S	I

(b) Mrs Jones chooses a meal at random.
What is the probability that she chooses

(i) fish followed by trifle,

(ii) a meal that includes steak pie,

(iii) a meal that does not include ice cream?

MEG

4 In a game, a three-sided spinner is spun and a dice is rolled.
The two numbers are **multiplied** to give a score.

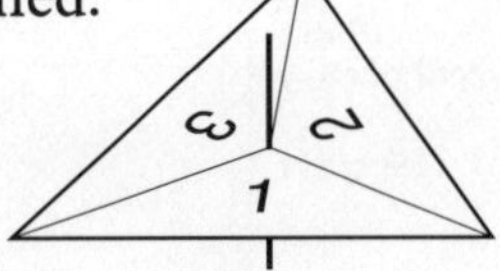

(a) Copy and complete the table to show all the possible scores.

(b) The spinner is spun once and the dice is rolled once.
What is the probability of:

(i) a score of 12,

(ii) a score of 5 or below,

(iii) a score of 8 or above?

	Dice					
Spinner	1	2	3	4	5	6
1		2				6
2						
3			9		15	

5 The diagram shows two sets of cards, grey and white.

One card is chosen at random from each set.

(a) List all the possible outcomes.

(b) What is the probability that both cards show the same number?

The two numbers are added together.

(c) What is the probability of getting a total of 5?

6 A red dice and a blue dice are both numbered 1 to 6.
In a game, both dice are thrown and the total score is found by adding the two numbers.

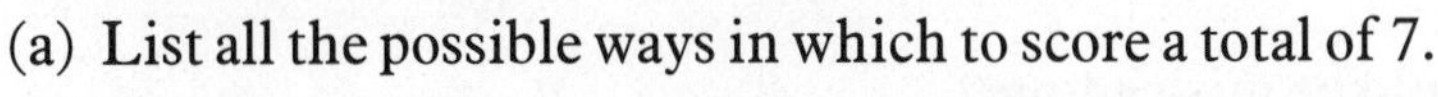

(a) List all the possible ways in which to score a total of 7.

(b) Explain which is more likely, a total score of 7 or a total score of 2.

Answers and hints ► page 141

Misleading graphs

There are a variety of ways that graphs can mislead.

Look carefully at the vertical and horizontal scales:
- Do they start at 0?
- Do numbers go up evenly?

Look carefully at any diagrams and pictures:
- Comparing parts, do they look too large or too small?

1 These two diagrams have been drawn to show what you need to eat for a healthy life.

A

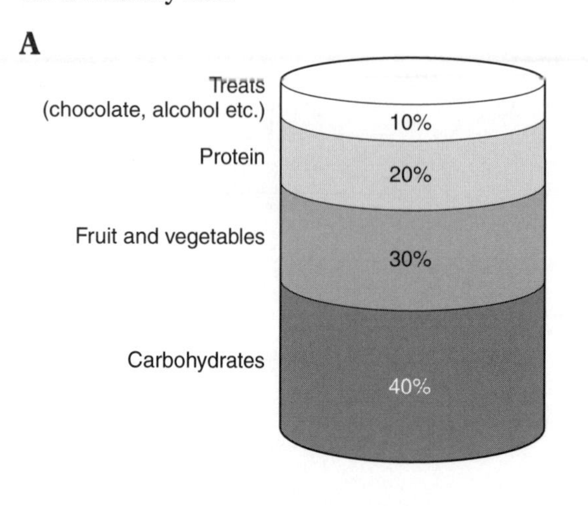

B

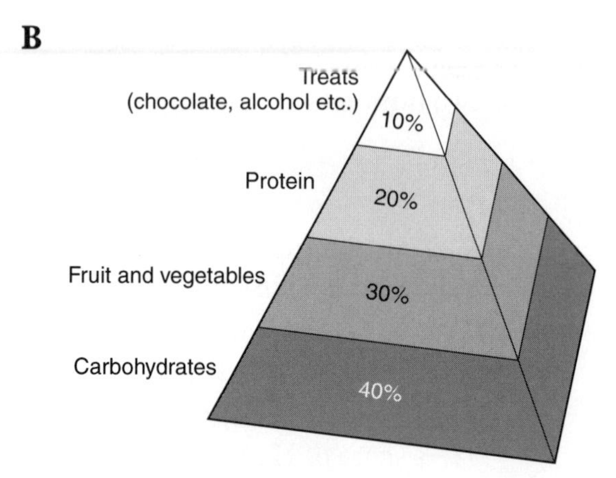

(a) Which diagram is misleading?

(b) Give a reason for your answer.

MEG (SMP)

2 These graphs show information about numbers of pupils at two schools.

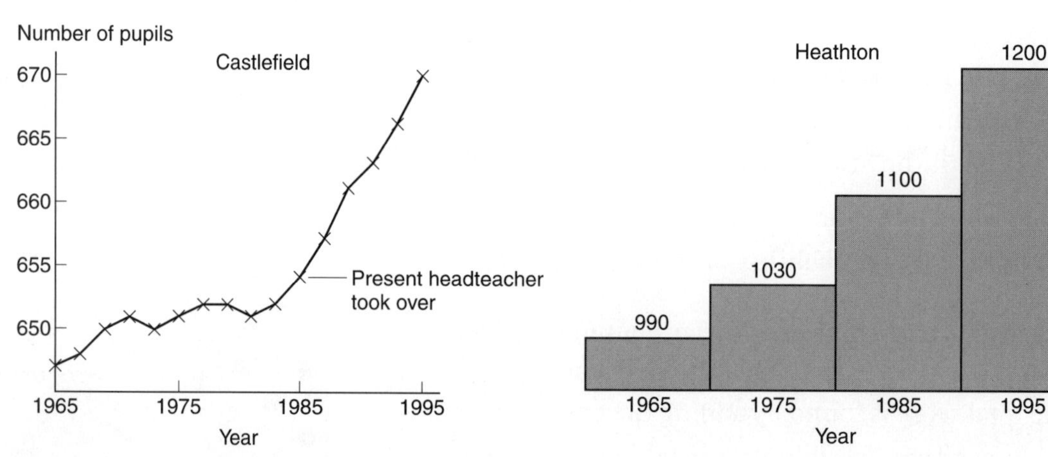

Describe how each graph could be misleading.

Answers and hints ► page 143

Mixed handling data

1 A shoe shop keeps a record of the number of pairs of trainers sold of each size. Here are the results for one month.

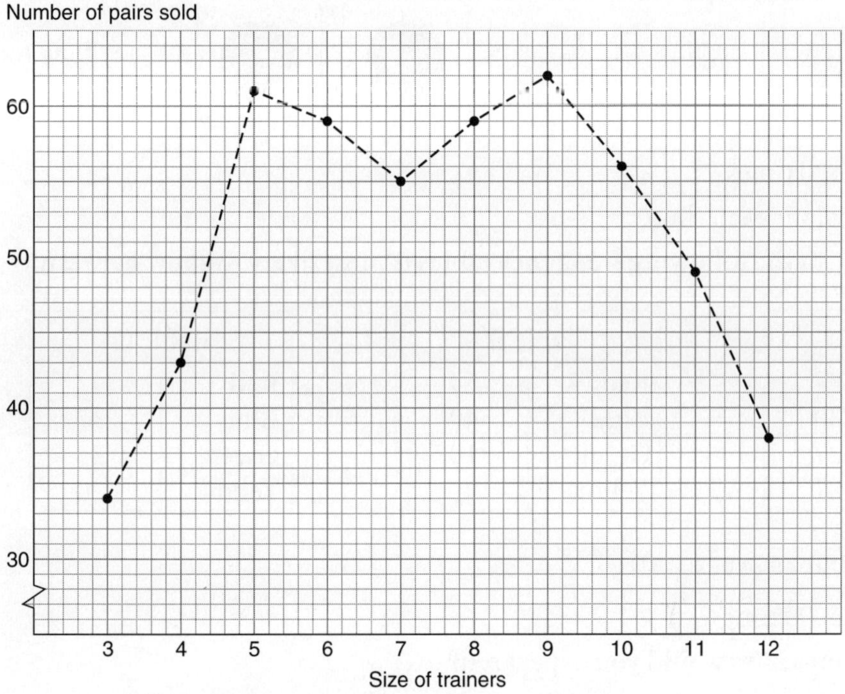

(a) (i) Which size of trainers was sold the most?

(ii) How many pairs was that?

(b) Which sizes of trainer sold less than 40 pairs?

(c) Explain why a jagged line (⦚) has been used on the vertical axis.

(d) Explain why the points have been joined up with dotted lines and not solid.

MEG (SMP)

2 The table gives details of flights on the Belfast to London route. Write down the missing entries.

Flight Number	Departure Time	Arrival Time	Flight Time
NI 207	0840	0936	(a)
NI 403	1310	(b)	1hr 12mins
NI 171	(c)	1744	1hr 15mins

NICCEA

3 This dice has a colour on each face.

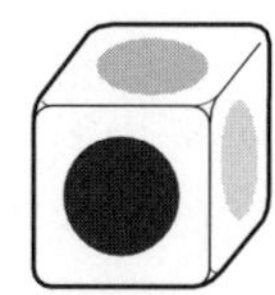

Its faces are shown below.

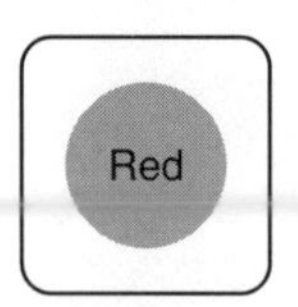

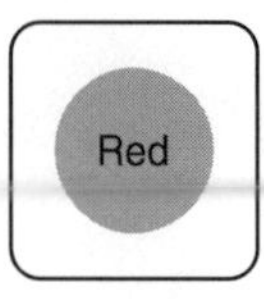

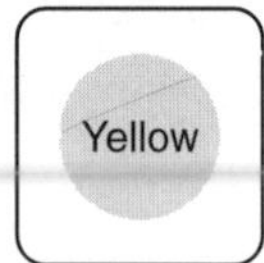

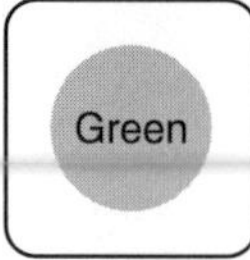

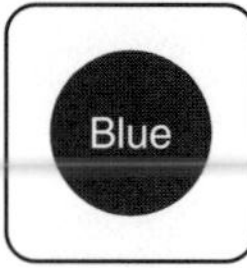

The dice is rolled 120 times.
Josie records the colour each time.

(a) (i) Which one of the following frequency tables is most likely to show her results?

Table A	
Red	32
Blue	30
Green	27
Yellow	31

Table B	
Red	64
Blue	40
Green	10
Yellow	6

Table C	
Red	42
Blue	37
Green	21
Yellow	20

(ii) Explain your answer to (a) (i).

The dice is rolled 6000 times.

(b) About how many times would you expect red?

4 These are the rules of a game at a summer fair.

Roll a 10p coin onto the board. If it lands in a square you win. If it lands on a line you lose.

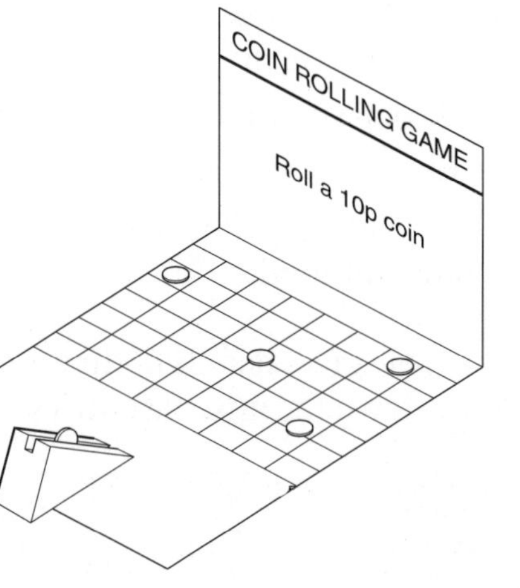

Jules tested the game by rolling a 10p coin 100 times.
These are his results.

	Tally marks	Frequency
Win	卌 卌 卌 III	
Lose	卌 卌 卌 卌 卌 卌 卌 卌 卌 卌 卌 II 卌 卌 卌 卌 卌	

(a) Copy and complete the frequency column.

(b) What is the experimental probability of a win?

(c) At the fair, 250 people play the game.
About how many people would you expect to win?

MEG (SMP)

5 Here are four frequency graphs. They show how the populations of four districts split up into different age groups.

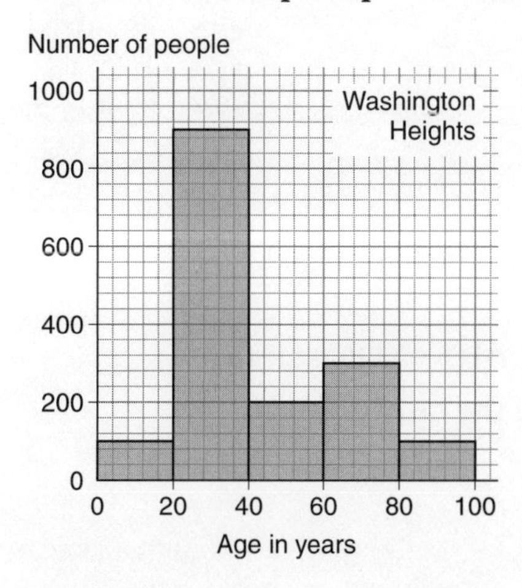

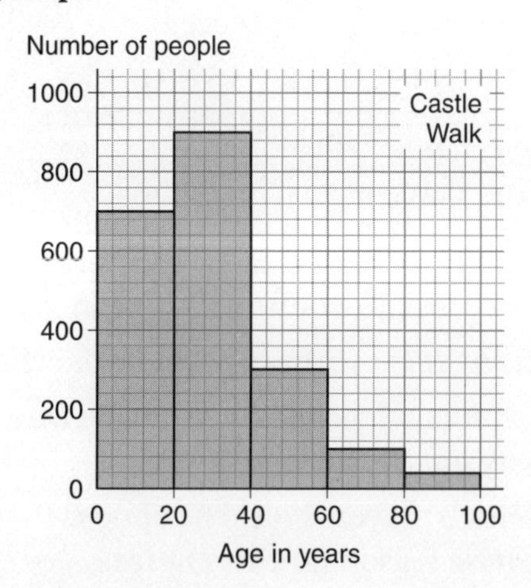

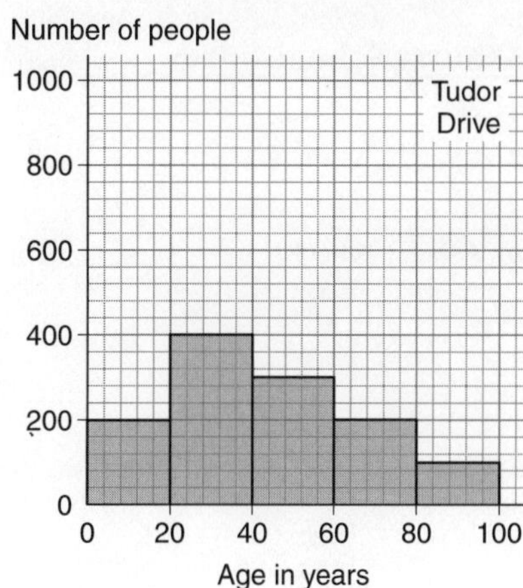

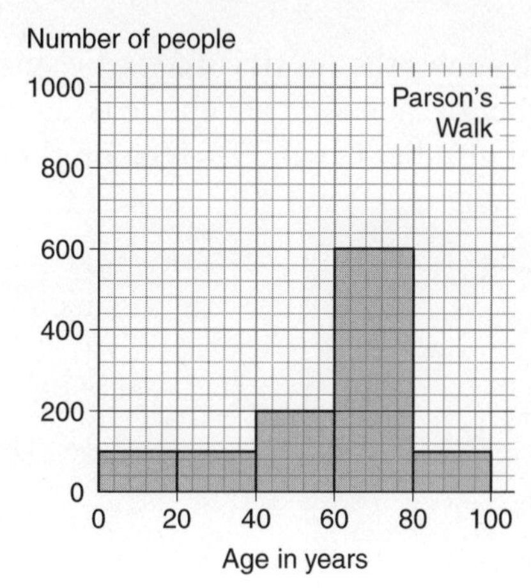

(a) (i) How many people live on Washington Heights?
(ii) Which district has the greatest number of under 20's?
(iii) Which district has the least number of over 80's?

(b) (i) Which district has a modal age group of '$60 \leq$ age < 80'?
(ii) Which is the modal age group for Tudor Drive?

(c) Which districts do these descriptions best fit?
(i) Few children but most people are between 20 and 40.
(ii) More than half the people are over 60.
(iii) Made up mainly of young families with children.

(d) What fraction of the people from Tudor Drive are over 40?

(e) A person is picked at random from Washington Heights. Estimate the probability that the person is:
(i) between 20 and 40 years old,
(ii) over 60 years old.

MEG/ULEAC (SMP)

6 Alan and Dharmesh have each played for their school basketball team.
Alan has played in 9 games.
Here are the points he has scored.

23 18 14 21 8 15 16 26 12

(a) Calculate Alan's mean score.

(b) Find the range of Alan's scores.

Dharmesh has scored a total of 133 points in the 7 games he has played.
His mean score is 19 points and his range is 18 points.

(c) For the next game the captain has to choose either Alan or Dharmesh.

(i) Explain why choosing the one who has scored the most points may not be the fairest way.

(ii) Which player should he choose?
Give a reason for this choice.

MEG/ULEAC (SMP)

7 The graph shows the results of a traffic survey outside Ashurst School.
The number of each type of vehicle is shown as a percentage
of the total number of vehicles seen on the day of the survey.

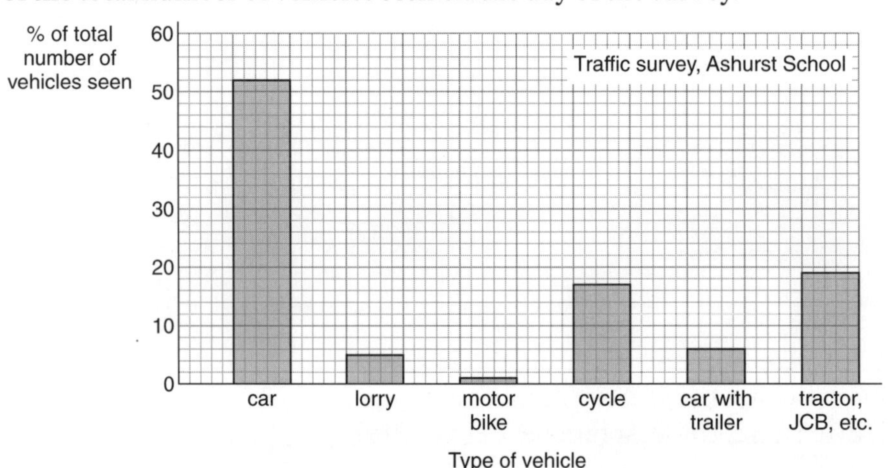

(a) Say whether each of these statements is TRUE or FALSE.

(i) The vehicle is most likely to be a car.

(ii) The vehicle is likely to be a motorbike.

(iii) The vehicle is unlikely to be a lorry.

Errol, from Ashurst School, said

'You would get very similar results outside any school.'

(b) Explain briefly how you could test whether this statement is true.

ULEAC

Answers and hints ► page 143

MIXED QUESTIONS

Mixed questions 1

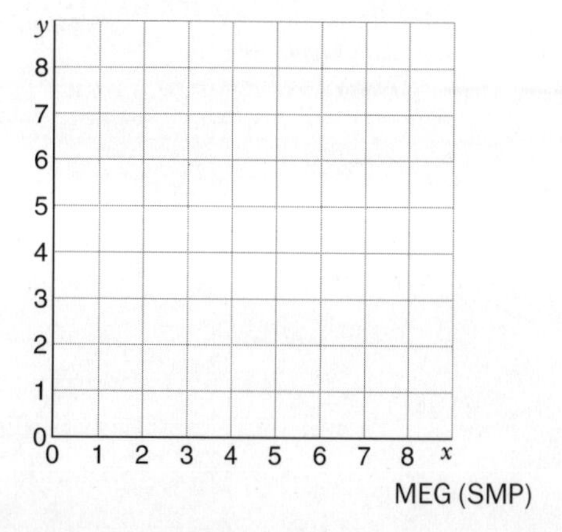

1 Copy the grid on to centimetre squared paper. Four corners of a pentagon are

A (4, 6), B (1, 3), C (1, 1) and D (7, 1).

(a) Plot and label A, B, C and D on the grid.

(b) The pentagon has one line of symmetry. Complete the drawing of the pentagon.

(c) Label the other corner E. Write down the coordinates of E.

(d) What is the equation of the line of symmetry?

MEG (SMP)

2 This table shows the cost of sending parcels by Parceltrip.

Weight **not over**	Delivery Time		
	Same day	Within 24 hrs	Within 48 hrs
10 kg	£15·45	£12·30	£9·50
11 kg	£16·25	£12·90	£9·95
12 kg	£17·05	£13·50	£10·40
13 kg	£17·85	£14·10	£10·85
14 kg	£18·65	£14·70	£11·30
15 kg	£19·45	£15·30	£11·75

(a) How much will it cost to send a parcel that weighs 11·5 kg for delivery on the same day?

(b) How much will this parcel cost to send for delivery within 24 hours?

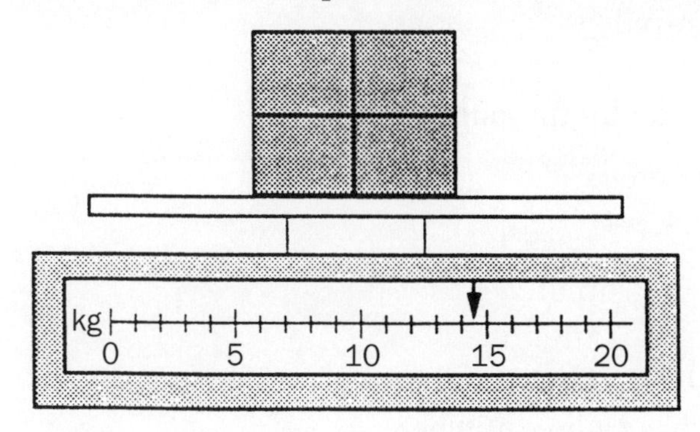

MEG (SMP)

3 Put these lengths in order, starting with the shortest.

1·63 m 1·12 m 1 m 8 cm 1·70 m 150 cm

4 York is 200 miles from Aberystwyth. The travel graph shows the journey of a coach travelling from Aberystwyth to York. It also shows the journey of a car travelling from York to Aberystwyth.

Distance from Aberystwyth (miles)

York 200

150

100

50

Aberystwyth 0

12 noon 2 4 6 8 10

Time (p.m.)

(a) At what time did the coach and the car pass each other?

(b) The coach stopped for two hours in Manchester.
Use the graph to say how far Manchester is from York.

(c) How long did the car journey take?

(d) The average speed of the car can be worked out using the formula

$$\text{average speed} = \frac{\text{distance travelled}}{\text{time taken}}.$$

Work out the average speed of the car for the journey.

MEG (SMP)

5 Here is a number machine chain.

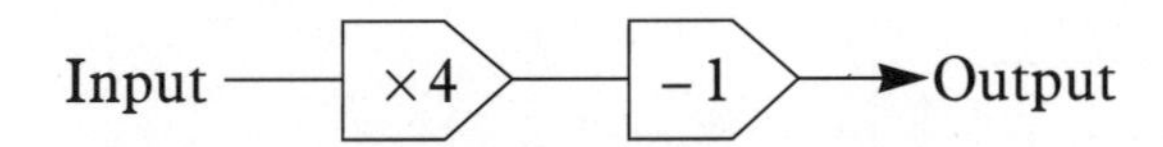

(a) The input is 2. What is the output?

(b) Paul says that for any input which is a whole number,
the output is always a prime number.
Give an input and its output which show that Paul is wrong.

(c) What is the output for an input of n?

MEG/ULEAC (SMP)

6 (a)

1	2	3	4	5	6	7	8	9	10
11	12	13	14	15	16	17	18	19	20
21	22	23	24	25	26	27	28	29	30
31	32	33	34	35	36	37	38	39	40

From the numbers shown above, write down

(i) the square of 6,

(ii) all of the multiples of 8,

(iii) the prime factors of 70,

(iv) the values of 3^3 and $\sqrt{25}$.

(b) (i) Write down the next two numbers in the number pattern

3, 7, 11, 15, 19, ..., ...

(ii) Write down in words the rule for finding the next number in the pattern from the one before it.

(iii) Write down an expression for the nth number in the pattern. WJEC

7

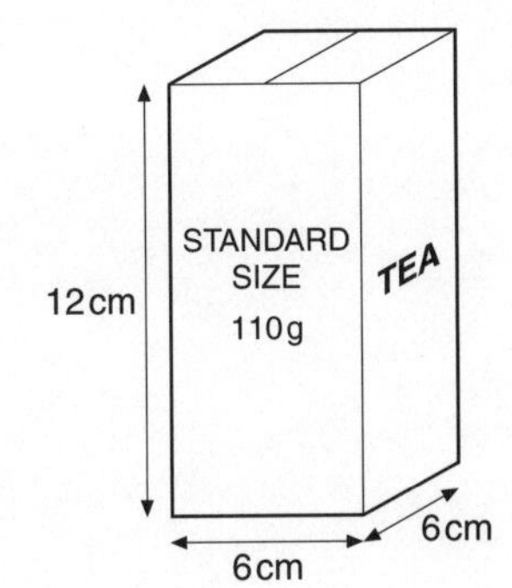

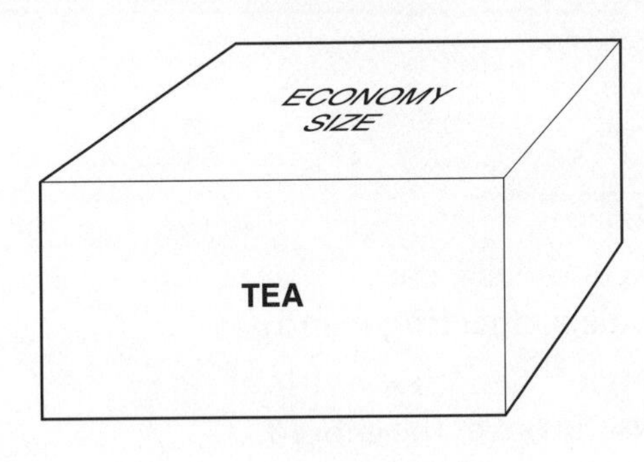

Not to scale

(a) Calculate the volume of the Standard Size packet of tea.

(b) The volume of the Economy Size packet is $1296\,\text{cm}^3$.
Calculate the weight of tea in the Economy Size packet, assuming both packets are completely filled. MEG

8 (a) One length of a swimming pool is 25 metres.
Susan swam 1 kilometre in a sponsored swim.
How many lengths did she swim?

*Do not use a calculator. Remember that you **must** show your working.*

(b) Cala swam 29 lengths.
She was sponsored at £1·95 per length.

(i) She wants to estimate how much she will collect.
Write down a calculation she could do in her head.

(ii) Explain how you know whether this estimate is bigger or smaller than the exact amount. MEG (SMP)

Answers and hints ► page 144

Mixed questions 2

1 This is a recipe for a curry sauce.
It serves 4 people.

(a) How much coconut milk do you need to serve 8 people?

(b) How many small aubergines do you need to serve 2 people?

(c) The sauce takes 15 minutes to cook.
You want it to be ready at 6:25 p.m.
At what time should you start to cook it?

THAI RED CURRY SAUCE

Serves four

1 tablespoon sunflower oil
2 tablespoons red curry paste
400g coconut milk
6 small aubergines

MEG (SMP)

2 In this question all measurements are in centimetres.
The volume of a cuboid = length × width × height.
A cuboid has edges d, 5 and 4.

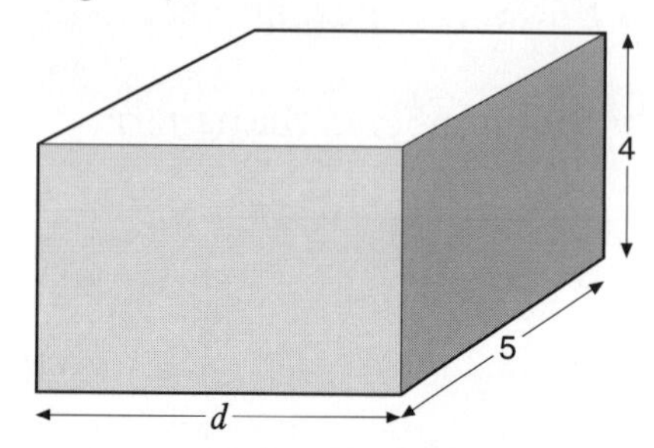

(a) The volume of the cuboid is $V\,\text{cm}^3$.
Write down a formula connecting V and d.

(b) The area of a rectangle = length × width.
The diagram shows the net of the cuboid.

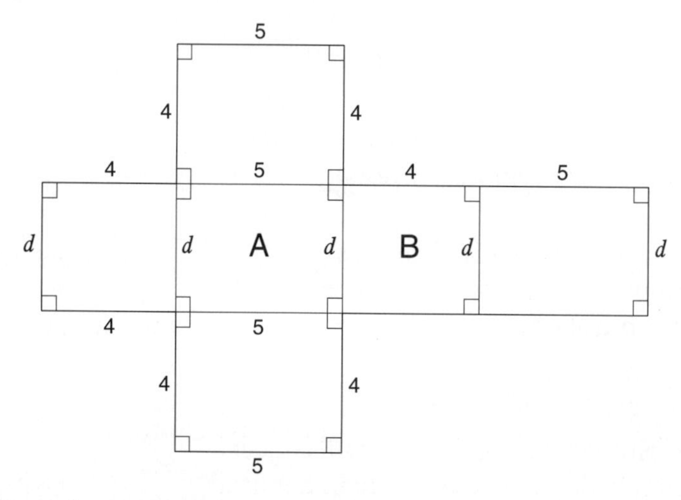

Write down, in terms of d,

(i) the area of rectangle A,

(ii) the area of rectangle B,

(iii) the total surface area of the cuboid.

MEG (SMP)

3 A shop sells adhesive edging.
Sara sold 3 pieces with these lengths:

$2\frac{1}{2}$ yards, $1\frac{3}{4}$ yards and $1\frac{1}{2}$ yards.

(a) How much did she sell altogether?

She cut this off a roll 10 yards long.

(b) How much was there left?

There are 36 inches in a yard.
Albert came in and asked for 120 inches.
Sara sold him $3\frac{1}{2}$ yards.

(c) How many inches did he have left over? MEG/ULEAC (SMP)

4 Simon travelled from Swansea to Cardiff by bus.
The distance from Swansea to Cardiff is 40 miles.

(a) The bus left Swansea at 11:34 and arrived at Cardiff at 13:25.
How many minutes did the journey take?

(b) Susan left Swansea at 12:00 and travelled to Cardiff by car.
The average speed for the journey was 30 m.p.h.
At what time did Susan arrive in Cardiff? WJEC

5 Mr and Mrs Jones buy a building plot for £20000.
They intend to build a house on it.
The plan of the ground floor of the house is shown below.

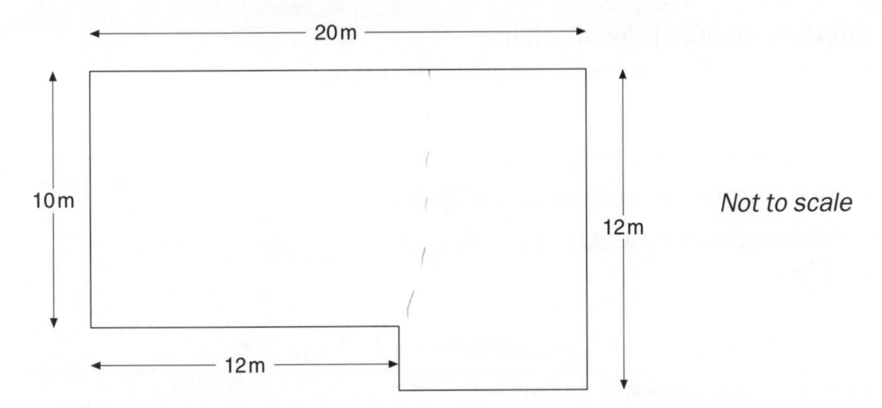

(a) Calculate the area of the ground floor.

(b) The cost of building the house is £225 for every square metre of ground floor.
Calculate the cost of building the house.

(c) Calculate the architect's fee, which is 6% of the cost of building the house. MEG

Answers and hints ► page 145

Mixed questions 3

1 The following formula can be used to find the cost, in pounds, of repairing a washing machine.

Cost in pounds $= 45 + 30 \times$ number of hours worked

(a) Calculate the cost of a repair when the work took half an hour.

(b) The cost of repairing a washing machine was £90.
How many hours did it take to repair the machine. WJEC

2

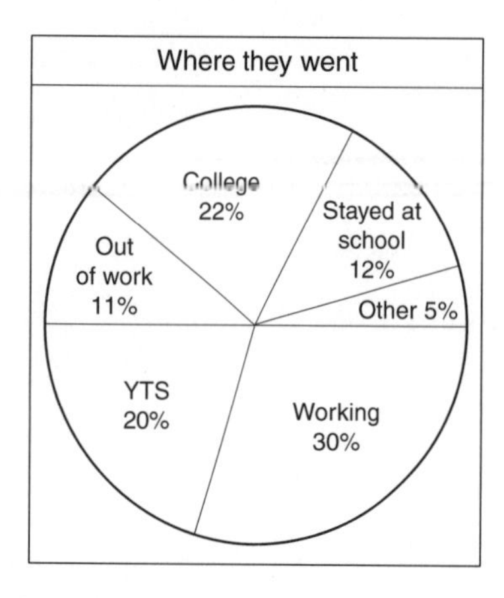

Not to scale

300 young people were asked what they did after completing Year 11 at school.
The pie chart shows the results of the survey.

(a) How many of the young people were working?

Gwen made an accurate drawing of the pie chart.
She first drew the sector representing the young people out of work.

(b) Calculate the size of the angle of this sector.
Give your answer correct to the nearest degree.

(c) Change to a decimal the percentage going to college.

(d) What fraction of the young people stayed at school?
Give your answer in its simplest form. ULEAC

3 On squared paper draw x- and y-axes with values of x and y from $^-5$ to 5.

(a) Plot the following points and join them up.

A (2, 4), B (5, 1), C (2, $^-2$), D ($^-2$, $^-2$), E ($^-5$, 1), F ($^-2$, 4)

(b) What is the name of shape ABCDEF?

(c) Mark one obtuse angle.

(d) What is the order of rotational symmetry of ABCDEF?

4 It takes 100 g of flour to make 15 shortbread biscuits.

(a) How many shortbread biscuits can be made from 1 kg of flour?

(b) Calculate the weight of flour needed to make 24 biscuits. MEG (SMP)

5

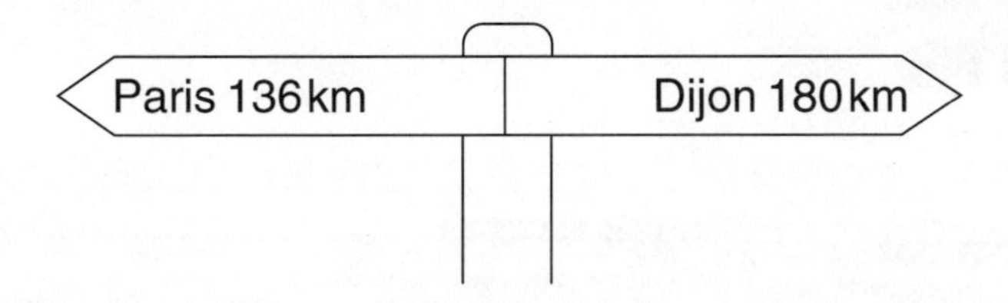

This signpost is on the road from Paris to Dijon.

(a) Work out the distance, in kilometres, from Paris to Dijon along this road.

(b) Work out the approximate distance, in miles, from the signpost to Paris.

Andrew drove to Dijon **starting from the signpost** at an average speed of 80km per hour.

(c) How long did the journey take?
Give your answer in hours and minutes. ULEAC

6 Keith buys a packet of marzipan for a birthday cake.
The packet measures 3·5cm by 6cm by 11cm.

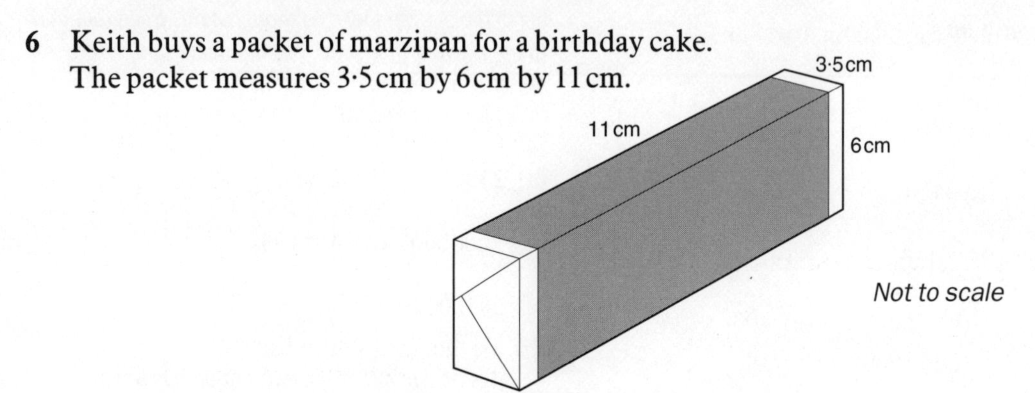

(a) What is the volume of the marzipan?

(b) The marzipan weighs 250g.
The birthday cake and marzipan together weigh 1·6kg.
What percentage of this weight is the marzipan?

(c) The top of the cake is a rectangle measuring 25cm by 20cm.

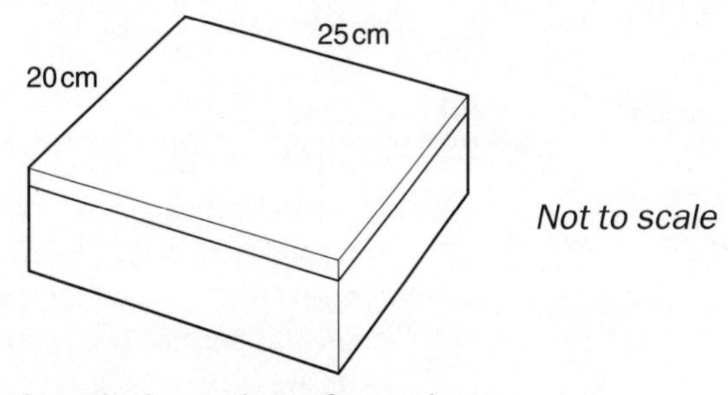

Keith uses the whole packet of marzipan.
How thick will the marzipan be when it is rolled out to cover the top of the cake? SEG

Answers and hints ► page 146

ANSWERS AND HINTS

NUMBER

Place value (page 2)

1 (a) 4 hundreds (b) 4 hundredths
(c) 4 thousands (d) 4 tenths
(e) 4 thousandths

2 (a) 4008 (b) 26120
(c) 310000 (d) 201000

3 988, 999, 1009, 1080, 1100

4 0·23 m, 0·234 m, 0·4 m, 0·409 m, 0·415 m

5 9·91 seconds
The winner is the runner with the smallest time.

6 (a) 1268 (b) 8621

7 (a) 5921 (b) 5129 (c) 5192

8 (a) 900 (b) 200 (c) 4000
(d) 1200 (e) 14000 (f) 100
(g) 80 (h) 5 (i) 500
(j) 500 (k) 6600 (l) 53
(m) 72000 (n) 8100 (o) 145

9 (a) 54 (b) 810 (c) 8
(d) 10·2 (e) 7 (f) 0·72
(g) 7·57 (h) 0·88 (i) 0·810
(j) 0·333
0·72 can be written as ·72 but the zero before the point makes it clearer.

10 (a) $\times$ (b) $\div$ (c) $\div$ (d) $\times$

11 Height in cm = $0{\cdot}6 \times 100 = 60$

12 Thickness in cm = $1{\cdot}2 \div 10 = 0{\cdot}12$

13 Multiplying by 10 makes the answer bigger.
But $3{\cdot}2 = 3{\cdot}20$, so the answer must be wrong.

More help or practice
Place value ► Book G4 pages 6 to 9
Multiplying and dividing by 10 and 100
► Book G5 pages 8 to 15, 26 to 30; Book B5 pages 91 to 92
Writing large numbers ► Book G8 pages 18 to 21
Multiplication patterns ► Book B4 pages 36 to 37

Rounding (page 4)

1 (a) 4800 (b) 72300 (c) 3000
(d) 4400 (e) 1000 (f) 3300
In (c) the answer looks as though it has been rounded to the nearest thousand, but the nearest hundred appears to be zero hundreds.
In (e) 9 is rounded up to 10 to give 1000.

2 (a) 25380 (b) 25000

3 (a) 5000000 (b) 4811000

4 (a) 283000
(b) To the nearest hundred
It is possible that the population has been rounded to the nearest 10. This would happen if the original figure had been, say, 279496.

5 14380, 14231, 14394, 14376 and 14351 cannot be winning entries.

6 (a) To the nearest million
(b) RECORD GATE OF 13000

7 (a) 2·48 (b) 9·31 (c) 5·01
(d) 4·28 (e) 0·40 (f) 0·70
In (e) the 9 is rounded to 10 which means that 0·3965 is rounded up to 0·40.
In (e) and (f) it is important to keep the zero in the hundredths position because it shows the accuracy of the answer.

8 (a) 2·2 (b) 7·6 (c) 0·1
(d) 11·9 (e) 5·0 (f) 6·0
In (e) 9 is rounded up to 10 which gives 5·0.
In (e) and (f) it is important to keep the zero in the answer.

9 (a) 5·9 (b) 5·89 (c) 5·890 (d) 5·8903

10 (a) £7·52 (b) £9·63 (c) £5·71
(d) £4·08 (e) £10·49 (f) £0·80

11 (a) 21·34 = 21·3 (to 1 d.p.)
(b) 18·5885 = 18·59 (to 2 d.p.)
(c) 60·869 … = 60·87 (to 2 d.p.)
(d) 5·185 … = 5·2 (to 1 d.p.)
(e) 7·466 = 7·47 (to 2 d.p.)
(f) 0·05818 … = 0·058 (to 3 d.p.)

12 £3·50 ÷ 6 = £0·5833333
= £0·58 or 58p (to the nearest penny)

13 63·9p × 46 = 2939·4p
= 2939p or £29·39 (to the nearest penny)

More help or practice
Rounding up and down ► Book G6 pages 28 to 32
Rounding large numbers ► Book G8 page 22, Book B1 pages 15 to 17
Rough estimates ► Book G6 pages 33 to 36
Rounding to a given number of decimal places ► Book G9 page 79, Book B1 pages 26 to 32
Rounding to the nearest penny ► Book B1 page 42

Pencil and paper calculations (page 6)

*Remember to show **all** your working. You will get marks for method even if you make a mistake in your calculations.*

1 (a) 932 (b) 775 (c) 332
(d) 512 (e) 179

2 (a) 1564 (b) 13653 (c) 124
(d) 33 (e) 34

3 53 – 28 = 25 (empty seats)

4 Tom scores (54 + 38 + 63) runs = 155 runs.
Richard scores (19 + 44 + 76) runs = 139 runs.
So Tom scores 155 – 139 = 16 runs more than Richard.

5 (a) 9 × 2 = 18 (hinges)
(b) 9 × 16 = 144 (screws)
Remember that it does not matter in which order you multiply numbers together.

6 (a) 96 ÷ 6 = 16 (tables)
(b) 96 ÷ 3 = 32 (benches)
You could also get the number of benches by multiplying the number of tables by 2 (16 × 2 = 32).

7
$$\begin{array}{r} \boxed{7}\ 5\ 8 \\ -\ 2\ \boxed{7}\ 3 \\ \hline 4\ 8\ \boxed{5} \\ \hline \end{array}$$

Remember to check by adding: 273 + 485 = 758.

8 (a) Sally spends 34p × 7 = 238p.
Sally's change should be
500p – 238p = 262p or £2·62.
Convert £5 to pence (500p) so you are working in the same units.
You would gain full marks for an answer in pence or pounds, but might lose a mark for £2·62p.
(b) Each CD costs £98 ÷ 8 = £12·25.

9 Number of crates = 840 ÷ 24 = 35

10 Total cost = 32p × 36 = 1152p = £11·52
You would lose a mark if you did not convert your answer from pence to pounds in this question.

11 (a) Alex will earn £65 × 12 = £780.
(b) She budgets to spend a total of £25 per week.
This means she can save £65 – £25 = £40 per week.
So in 12 weeks Alex saves £40 × 12 = £480.

12 Number of rows of chairs = 368 ÷ 16 = 23.

13 (a) *The operation that reverses multiplying by 2 is dividing by 2.*
If you divide 58 by 2, you get the number Emma is thinking of.
58 ÷ 2 = 29
So Emma is thinking of 29.
(b) *The operation that reverses adding 23 is subtracting 23.*
71 – 23 = 48.
So Aziz is thinking of 48.

14 The number of boys is $912 - 474 = 438$.
$474 - 438 = 36$
There are 36 more girls than boys.

15 (a) The number of bottles $= 252 \div 14 = 18$
(b) John's change in francs $= 300 - 252 = 48$

16 $£50 = 50 \times 420$ drachmas
$= 21\,000$ drachmas
The bank charges a fee of 800 drachmas.
So Rashid will get
$21\,000 - 800$ drachmas $= 20\,200$ drachmas.

> **More help or practice**
> Methods of multiplication ► Book G+ pages 17 to 22, Book B+ pages 8 to 14
> Multiplication problems ► Book G9 page 13, Book G+ page 23
> Methods of division ► Book G+ pages 66 to 67, Book B+ pages 43 to 48
> Division problems ► Book G9 page 33
> Mixed problems ► Book G5 pages 1 to 6

Rough estimates (page 8)

1 $12 + 4 + 9 = 25$
Pete has rounded each price up, so he **can** afford the clothes.

2 (a) 600×70 is the best approximation.
(b) The approximate answer would be too large, since both numbers have been rounded up.

3 $£2{\cdot}89 \times 210$ is roughly equal to $£3 \times 200 = £600$.

4 (a) $£4 \times 2 + £2 = £8 + £2 = £10$
(b) $£7{\cdot}99 \times 6$ is roughly equal to $£8 \times 6 = £48$.
So £50 is the amount which is closest.

5 4 hours 45 minutes is roughly equal to 5 hours.
So Sushma earns roughly $£20 \div 5 = £4$ per hour.

6 (a) $£10 \times 50 = £500$
(b) Both numbers are rounded up, so the estimate is bigger than the exact cost.

7 Amy will receive roughly
60×200 pesetas $= 12000$ pesetas.

> **More help or practice**
> Rough estimates ► Book B4 pages 49 to 51

Negative numbers (page 9)

1 $^-7$°C, $^-2$°C, 0°C, 3°C, 4°C

2 (a) The temperature inside the tent was $^-1$°C.
(b) The temperature had gone up 9°C.
(c) $^-4$°C, $^-2$°C, $^-1$°C

3 (a) Aberdeen had the lowest temperature.
(b) Plymouth had the highest temperature.
(c) The temperature in Dublin was $^-2$°C.
This is the same temperature as in Glasgow.

4 The difference is 270°C.

> **More help or practice**
> Negative numbers ► Book G3 page 28, Book G+ pages 59 to 63, Book B1 pages 83 to 87, Book B+ pages 56 to 59

Using a calculator (page 10)

1 *For each part a suggested rough estimate is given in brackets.*

(a) 6·3 $[\frac{10 + 10}{3} = \frac{20}{3} = \text{nearly } 7]$
First work out the addition and then divide by 2·7.

(b) 0·61 $[\frac{60 - 30}{50} = \frac{30}{50} = 0{\cdot}6]$
First work out the subtraction and then divide by 50.

(c) 0·8 $[\frac{1 \times 1}{2} = \frac{1}{2} = 0{\cdot}5]$
$1{\cdot}2^2 = 1{\cdot}2 \times 1{\cdot}2$.

(d) 26·67 $[4 \times (4 + 3) = 28]$
First work out the addition and then multiply by 4·2.

(e) 21·12 $[(20 \times 1) - 4 = 16]$
First work out the multiplication and then subtract 3·6.

(f) 9·7 $[8 + \sqrt{4} = 8 + 2 = 10]$
First calculate the square root and then add 7·8.

2 (a) Sabrina pays £4·99 × 2·1 = £10·479
= £10·48 (to the nearest penny)

(b) A rough check would be 5 × 2 = 10.

3 (a) 2·52 + 4·19 + 1·87 = 8·58
Jacinta walked 8·58 km.

(b) 2·52 – 1·87 = 0·65
She walked 0·65 km further in the morning.

4 (a) 10·8625 = 10·9 (to 1 d.p.)

(b) 8·660254 = 8·7 (to 1 d.p.)

(c) 5·832 = 5·8 (to 1 d.p.)
Hint: $1{\cdot}8^3$ means 1·8 × 1·8 × 1·8.

5 86·4 kg – 62·9 kg = 23·5 kg

6 *Remember that you must use the same units of money: pounds or pence. Here are two ways of doing the calculation.*

Number of litres = 2500 ÷ 69·5 (£25 = 2500p)
= 35·971 …
= 36 (to the nearest litre)

Number of litres = 25 ÷ 0·695 (69·5p = £0·695)
= 35·971 …
= 36 (to the nearest litre)

7 The coke costs £0·45 × 3 = £1·35.
The coke and lemonade cost
£1·35 + £1·12 = £2·47.
Laura's change = £5·00 – £2·47 = £2·53.

8 (a) The total number of bookings = 3854

(b) The total = 3900 (to the nearest hundred)

9 Total cost = (£0·18 × 450) + (£42 × 2)
= £81 + £84
= £165

10 (a) Cost per leaflet = £4 ÷ 200 = £0·02 or 2p

(b) Number of leaflets delivered = 7 ÷ 0·02
= 350

You could also work in pence:
400p ÷ 200 = 2p and 700 ÷ 2 = 350

11 Weekly wage = £4·22 × 28
= £118·16

12 *The best way to answer this question is to find the total cost for one person and then multiply by 4.*
£18·75 + £2·30 = £21·05
£21·05 × 4 = £84·20

13 Total paid in instalments = £24·50 × 36
= £882

Total paid altogether = £882 + £100
= £982

The difference = £982 – £699
= £283

Jahal would have saved £283 by paying cash.

14 Basic pay = £6·26 × 36 = £225·36

3 hours overtime at time and a half means that Mrs Kahn is paid for $3 \times 1\frac{1}{2}$ hours = $4\frac{1}{2}$ hours.
4 hours overtime at double time means that she is paid for 4 × 2 hours = 8 hours.

Total overtime pay = £6·26 × 12·5 = £78·25
Total pay = £225·36 + £78·25
= £303·61

More help or practice

Using a calculator ► Book G1 pages 52 to 56; Book G2 pages 40 to 47; Book B2 pages 29 to 33, 80 to 84; Book B4 pages 1 to 5

Checking by rounding ► Book B1 pages 46 to 49

Calculations with decimals ► Book G2 pages 36 to 38; Book B1 pages 40 to 44, 52 to 55, 58 to 59

Comparing prices by working out unit costs ► Book B1 pages 55 to 57

Leftovers (page 12)

1 $75 \div 6 = 12$ remainder 3 or 12·5
Latoya needs to buy 13 packs.
You need to round up so that there are enough invitations for everyone.
This means 3 invitations will be left over.

2 (a) $500 \div 19 = 26{\cdot}315\ldots$ *£5 = 500p*
Jenny buys 26 stamps.
You need to round down because you can't buy part of a stamp.
(b) The stamps cost $19\text{p} \times 26 = 494\text{p}$,
so she gets $500\text{p} - 494\text{p} = 6\text{p}$ change.

3 $550 \div 17 = 32{\cdot}352\ldots$
Least number of trips = 33
Here you need to round up; the cable car will not be full on the last trip.

4 (a) $40 \div 12 = 3{\cdot}333\ldots$
Alan needs to buy 4 packs.
(b) He pays $£1{\cdot}89 \times 4 = £7{\cdot}56$

5 (a) Thickness $= 163{\cdot}8\text{cm} \div 13$
$= 12{\cdot}6\text{cm}$
(b) $100 \div 12{\cdot}6 = 7{\cdot}936\ldots$
There will only be room for 7 shoe boxes.
The eighth won't quite fit!

6 (a) $209 \div 24 = 8{\cdot}708\ldots$
The least number of rows needed is 9.
(b) The number of seats in 9 rows is
$24 \times 9 = 216$.
The number of spare seats will be
$216 - 209 = 7$.

More help or practice
Division: rounding up and down and remainders
► Book G+ pages 41 to 46

Trial and improvement (page 13)

You must show all your trials. You may get some method marks even if your final answer is wrong.
In the answers we have shown one set of trials. You may have used different numbers or needed a few more trials.

1

	Width tried	Working		
	5cm	$5 \times 5 = 25$	Too small	
	7cm	$7 \times 7 = 49$	Too large	The solution is between 6 and 7.
(a)	6cm	$6 \times 6 = 36$	Too small	
(b)	6·5cm	$6{\cdot}5 \times 6{\cdot}5 = 42{\cdot}25$	Too large	
	6·2cm	$6{\cdot}2 \times 6{\cdot}2 = 38{\cdot}44$	Too small	
	6·3cm	$6{\cdot}3 \times 6{\cdot}3 = 39{\cdot}69$	Too small	The solution is between 6·3 and 6·4.
	6·4cm	$6{\cdot}4 \times 6{\cdot}4 = 40{\cdot}96$	Too large	

(c) $6{\cdot}3 \times 6{\cdot}3 = 39{\cdot}69$ is nearer to 40 than
$6{\cdot}4 \times 6{\cdot}4 = 40{\cdot}96$.
So the width of the card should be 6·3cm.

2

Trial number	Working	
7	$7 \times 7 = 49$	Too small
8	$8 \times 8 = 64$	Too large
7·7	$7{\cdot}7 \times 7{\cdot}7 = 59{\cdot}29$	Too small
7·8	$7{\cdot}8 \times 7{\cdot}8 = 60{\cdot}84$	Too small
7·9	$7{\cdot}9 \times 7{\cdot}9 = 62{\cdot}41$	Too large

The square root of 61 is between 7·8 and 7·9, but 60·84 ($= 7{\cdot}8^2$) is much closer to 61 than 62·41 ($= 7{\cdot}9^2$).
So $\sqrt{61} = 7{\cdot}8$ (to 1 d.p.)

3 (a) Area $= 4 \times 6 = 24\text{m}^2$
(b) Area $= 5 \times 7 = 35\text{m}^2$
(c) *The width must be between 4m and 5m, so start by trying, say, 4·6m.*

Width tried	Working	
4·6	$4{\cdot}6 \times 6{\cdot}6 = 30{\cdot}36$	Too small
4·7	$4{\cdot}7 \times 6{\cdot}7 = 31{\cdot}49$	Too small
4·8	$4{\cdot}8 \times 6{\cdot}8 = 32{\cdot}64$	Too large

31·49 is closer to 32 than 32·64.
So the width must be 4·7m (to 1 d.p.).

More help or practice
Trial and improvement ► Book G9 pages 85 to 89

Fractions (page 14)

1 $\frac{2}{6}, \frac{3}{9}, \frac{4}{12}, \frac{5}{15}, \frac{6}{18}, \frac{7}{21}, \frac{8}{24}, \frac{9}{27}, \frac{10}{30}, \ldots$

are all equal to $\frac{1}{3}$.

2 (a) *You will get full marks for shading **any** 16 squares.*

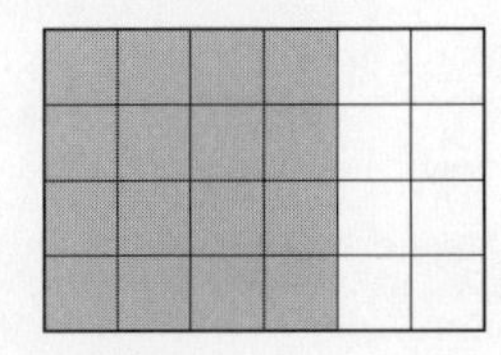

(b) (i) $\frac{20}{24} = \frac{5}{6}$ (ii) $\frac{4}{24} = \frac{1}{6}$

3 (a) $\frac{1}{8} + \frac{3}{4} = \frac{1}{8} + \frac{6}{8} = \frac{7}{8}$

(b) $\frac{7}{8} - \frac{3}{16} = \frac{14}{16} - \frac{3}{16} = \frac{11}{16}$

(c) $1\frac{1}{2} + \frac{3}{8} = 1 + \frac{4}{8} + \frac{3}{8} = 1 + \frac{7}{8} = 1\frac{7}{8}$

(d) $\frac{1}{2} - \frac{3}{16} = \frac{8}{16} - \frac{3}{16} = \frac{5}{16}$

(e) $2\frac{3}{8} + 1\frac{1}{4} = 2 + 1 + \frac{3}{8} + \frac{2}{8} = 3\frac{5}{8}$

Add the whole numbers separately.

4 A is $1\frac{3}{16}$, B is $1\frac{3}{8}$, C is $1\frac{9}{16}$, D is $1\frac{3}{4}$.

5 (a) $\frac{1}{2}$ of £5 = £5 ÷ 2 = £2·50

(b) $\frac{1}{3}$ of 24cm = 24cm ÷ 3 = 8cm

(c) $\frac{1}{5}$ of £35 = £35 ÷ 5 = £7, so

$\frac{2}{5}$ of £35 = £7 × 2 = £14

(d) $\frac{1}{4}$ of 500g = 500g ÷ 4 = 125g, so

$\frac{3}{4}$ of 500g = 125g × 3 = 375g

6 (a) Total weight $= (1\frac{1}{4} + 1 + \frac{3}{4} + \frac{1}{2} + 1\frac{1}{2})$lb

= 5lb

(b) (i) $\frac{1}{2}$lb + $\frac{1}{4}$lb = $\frac{3}{4}$lb

(ii) 5lb − $\frac{3}{4}$lb = $4\frac{1}{4}$lb

7 $\frac{15}{40}$ pupils swim more than 1 mile.

$\frac{15}{40} = \frac{3}{8}$ in its simplest form.

8 (a) $\frac{5}{20} = \frac{1}{4}$

(b) $\frac{1}{5}$ of 20 = 20 ÷ 5 = 4

Sally has eaten 4 pieces.

9 (a) $\frac{2}{16} = \frac{1}{8}$ (b) $\frac{12}{16} = \frac{3}{4}$ (c) $2\frac{8}{16} = 2\frac{1}{2}$

10 $\frac{1}{9}$ of 225g = 225g ÷ 9 = 25g

$\frac{4}{9}$ of 225g = 25g × 4 = 100g

100g of wood is cut away.

More help or practice
Equivalent fractions and simplest form
► Book B5 pages 84 to 87
Adding and subtracting fractions ► Book G7 pages 8 to 12
Finding a fraction of a quantity ► Book G8 page 35,
Book B2 page 83

Fractions, decimals and percentages (page 16)

1 (a) $25\% = \frac{25}{100} (= \frac{1}{4})$ (b) $30\% = \frac{30}{100} (= \frac{3}{10})$

(c) $35\% = \frac{35}{100} (= \frac{7}{20})$ (d) $68\% = \frac{68}{100} (= \frac{17}{25})$

(e) $21\% = \frac{21}{100}$ (f) $4\% = \frac{4}{100} (= \frac{1}{25})$

2 (a) $\frac{1}{2}$ = 1 ÷ 2 = 0·5

(b) $\frac{3}{4}$ = 3 ÷ 4 = 0·75

(c) $\frac{7}{10}$ = 7 ÷ 10 = 0·7

(d) $\frac{3}{8}$ = 3 ÷ 8 = 0·375

(e) $\frac{17}{100}$ = 17 ÷ 100 = 0·17

(f) $\frac{9}{100}$ = 9 ÷ 100 = 0·09

3 (a) $0{\cdot}8 = 0{\cdot}8 \times 100\% = 80\%$
(b) $0{\cdot}71 = 0{\cdot}71 \times 100\% = 71\%$
(c) $0{\cdot}32 = 0{\cdot}32 \times 100\% = 32\%$
(d) $0{\cdot}07 = 0{\cdot}07 \times 100\% = 7\%$
(e) $0{\cdot}025 = 0{\cdot}025 \times 100\% = 2{\cdot}5\%$
(f) $0{\cdot}375 = 0{\cdot}375 \times 100\% = 37{\cdot}5\%$

4 (a) $30\% = \frac{30}{100} = \frac{3}{10}$
(b) $0{\cdot}7 = \frac{7}{10}$
(c) $25\% = \frac{25}{100} = \frac{1}{4}$
(d) $0{\cdot}21 = \frac{21}{100}$
(e) $0{\cdot}03 = \frac{3}{100}$
(f) $74\% = \frac{74}{100} = \frac{37}{50}$

5 (a) $\frac{1}{4} = 1 \div 4 = 0{\cdot}25 = 25\%$
(b) $\frac{1}{5} = 1 \div 5 = 0{\cdot}2 = 20\%$
(c) $\frac{3}{4} = 3 \div 4 = 0{\cdot}75 = 75\%$
(d) $\frac{1}{8} = 1 \div 8 = 0{\cdot}125 = 12{\cdot}5\%$ or $12\frac{1}{2}\%$
(e) $\frac{36}{100} = 36 \div 100 = 0{\cdot}36 = 36\%$
(f) $\frac{9}{100} = 9 \div 100 = 0{\cdot}09 = 9\%$

6 (a) $\frac{7}{8} = 7 \div 8 = 0{\cdot}875$
(b) $0{\cdot}875 = 0{\cdot}875 \times 100\% = 87{\cdot}5\%$

7 (a) $\frac{12}{15} = 12 \div 15 = 0{\cdot}8$
(b) $0{\cdot}8 \times 100\% = 80\%$

8 *The easiest way to put the numbers in order is to change all the numbers to decimals.*

$0{\cdot}3$ $\quad \frac{1}{2} = 0{\cdot}5$ $\quad 20\% = 0{\cdot}2$

$\frac{1}{4} = 0{\cdot}25$ $\quad \frac{1}{3} = 0{\cdot}333\ldots$ $\quad 24\% = 0{\cdot}24$

So the order, smallest first, is
$20\%,\ 24\%,\ \frac{1}{4},\ 0{\cdot}3,\ \frac{1}{3},\ \frac{1}{2}.$

9 $12\% = \frac{12}{100} = 0{\cdot}12$
$£350000 \times 0{\cdot}12 = £42000$

10 $\frac{1}{8}$ of $£1360 = £1360 \div 8 = £170$
$\frac{3}{8}$ of $£1360 = £170 \times 3 = £510$

11 $£11400 \times 0{\cdot}04 = £456$
Remember that 4% = 0·04, not 0·4.

12 (a) $25\% = 0{\cdot}25$ and $500 \times 0{\cdot}25 = 125$
(b) $\frac{1}{10}$ of $500 = 50$
So $\frac{7}{10}$ of $500 = 350$

13 (a) $\frac{2}{10} = \frac{1}{5} = \frac{1}{5} \times 100\% = 20\%$
or $\frac{2}{10} = \frac{20}{100} = 20\%$
(b) $60\% = \frac{60}{100} = 0{\cdot}6$
60% of 9000 million $= 9000 \times 0{\cdot}6$ million
$= 5400$ million

14 $17{\cdot}5\% = 0{\cdot}175$
VAT $= £27{\cdot}50 \times 0{\cdot}175$
$= £4{\cdot}8125$
$= £4{\cdot}81$ (to the nearest penny)

More help or practice
Converting between fractions, decimals and percentages
► Book G5 pages 36 to 37, Book G6 pages 7 to 13, Book B2 pages 88 to 92
Calculating a percentage of a quantity
► Book G3 pages 8 to 13, Book G5 pages 38 to 41, Book B3 pages 39 to 42, Book B4 pages 44 to 45, Book B5 pages 22 to 23

Percentages (page 18)

1 (a) $15\% = \frac{15}{100} = 0{\cdot}15$
(b) $£350 \times 0{\cdot}15 = £52{\cdot}50$
(c) $£350 - £52{\cdot}50 = £297{\cdot}50$

2 (a) 7% of $£135 = £135 \times 0{\cdot}07$
$= £9{\cdot}45$
(b) $£135 + £9{\cdot}45 = £144{\cdot}45$

3 $\frac{300}{800} = 0{\cdot}375 = 0{\cdot}375 \times 100\% = 37{\cdot}5\%$
First work out the fraction of the block that is removed; write it as a decimal and multiply by 100% to change to a percentage.

4 (a) $60\text{cm} \times 0{\cdot}15 = 9\text{cm}$
The plant's new height is 69cm.

(b) The difference in height is 0·2m
Start by writing this as a fraction of the original height.
$\frac{0{\cdot}2}{1{\cdot}6} = 0{\cdot}2 \div 1{\cdot}6 = 0{\cdot}125$
Multiply by 100% to change the decimal to a percentage.
$0{\cdot}125 \times 100\% = 12{\cdot}5\%$

5 $\frac{3}{13} \times 100\% = 23{\cdot}076\ldots\%$
$= 23\%$ (to nearest whole number)

6 The total number of people employed is
$348 + 252 = 600$.
The fraction that are women is $\frac{348}{600} = 0{\cdot}58$.
The percentage is $0{\cdot}58 \times 100\% = 58\%$.

7 $35\% = \frac{35}{100} = 0{\cdot}35$
$340\text{g} \times 0{\cdot}35 = 119\text{g}$

8 $18\% = 0{\cdot}18$
The reduction in price is
£69·50 × 0·18 = £12·51.
So the sale price is £69·50 – £12·51 = £56·99.

9 *Start by writing the decrease as a fraction of the original number. Then convert the fraction to a percentage.*

$\frac{77}{1100} = 77 \div 1100 = 0{\cdot}07 = 7\%$

10 *Gareth*
7% of £645 = £645 × 0·07 = £45.15
Gareth's pay after the increase is
£645 + £45·15 = £690·15.

Vicky
5% of £670 = £670 × 0·05 = £33·50
Vicky's pay after the increase is
£670 + £33·50 = £703·50.

So Vicky had the higher pay after the increases.

11 The reduction at *Best Buys* is given by
£320 × 0·15 = £48.
So *Best Buys* will sell the midi system at
£320 – £48 = £272.
This price is the cheaper of the two.
The difference in price is
£289·99 – £272 = £17·99.

More help or practice

One quantity as a percentage of another
► Book B2 pages 93 to 94, Book B4 pages 46 to 47, Book B5 pages 24 to 25

Percentage increases and decreases
► Book G4 pages 52 to 57; Book B3 page 44; Book B5 pages 23 to 24, 26 to 28

VAT, saving and borrowing, hire purchase
► Book G8 pages 48 to 51; Book B3 page 43; Book B5 pages 50 to 51, 72 to 73, 88 to 90

Profit and loss ► Book G8 page 52

Wages ► Book G3 pages 53 to 55

Properties of numbers (page 20)

1 (a) 8 and 12 are multiples of 4.

(b) 8 is the square root of 64.

(c) 11 is the prime number.

2 (a) $5 \times 5 = 25$ (b) $2 \times 2 \times 2 = 8$

(c) $\sqrt{9} = 3$

3 (a) 3, 5 and 9 are odd numbers.

(b) 4 and 9 are square numbers.

(c) 3, 6 and 9 are multiples of 3.

4 $3^2 = 9$, $8^2 = 64$, $3^3 = 27$, $4^2 = 16$,
$5^3 = 125$, $2^2 = 4$
In order, smallest first, the numbers are
2^2, 3^2, 4^2, 3^3, 8^2, 5^3.

5 (a) (i) 5, 10 and 15 are multiples of 5.
(ii) 2, 3 and 5 are prime numbers.

(b) 3 is a factor of 3, 12 and 15.
3 divides exactly into these numbers.

6 (a) 6 is the smallest number which is a multiple of 2 and 3.
The next multiples of 2 and 3 are 12, 18, 24, ...

(b) 1, 2, 5 and 10 are factors of 10.

7 (a) 2, 3, 5 and 7 are the prime numbers less than 10.

(b) 1, 4, 9 and 16 are the square numbers less than 20.

(c) 8 is the only cube number between 5 and 10.
$1^3 = 1$, $2^3 = 8$, $3^3 = 27$, ...

8 Here are two factor trees. There are others.

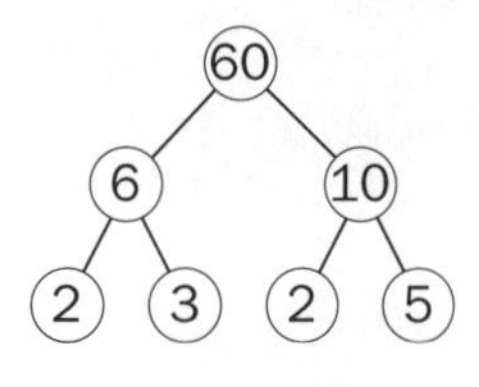

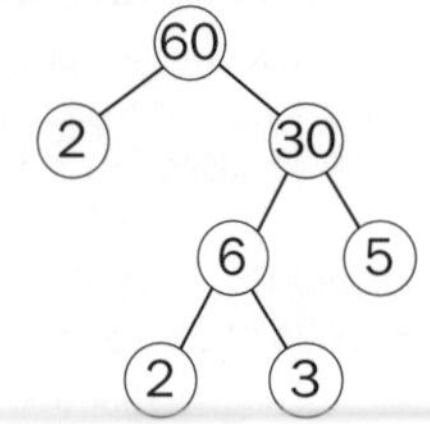

$60 = 2 \times 3 \times 2 \times 5$
$= 2^2 \times 3 \times 5$

$60 = 2 \times 2 \times 3 \times 5$
$= 2^2 \times 3 \times 5$

You need to continue the 'tree' until all the factors are prime numbers.

9 (a) 3, 13 and 23 are prime numbers.
(b) 4 and 25 are factors of 200.
(c) 28 is a multiple of 7.
(d) 64 is a cube number.
(e) $5 \div 15 = \frac{1}{3}$

10 (a) (i) 16 is a square number.
(ii) 18 is a multiple of 9.
(b) (i) $5 \times 5 = 25$ is the next square number.
(ii) 27 is the next multiple of 9.

More help or practice
Factors, multiples and prime numbers
► Book G9 pages 95 to 98, Book B+ pages 66 to 67
Squares and square roots ► Book G9 pages 99 to 100, 113; Book B5 page 70
Cube numbers ► Book G9 pages 101, 114

Ratio (page 22)

1 (a) 25 g (b) 24 toffee apples

2 0·2 litre

3 (a) 12·5 ounces or $12\frac{1}{2}$ ounces
(b) 36 mince pies
(c) 20 : 25 which simplifies to 4 : 5.

4 *Divide the money into 5 equal parts.*
(a) Jane receives 3 parts out of 5.
1 part = £600 ÷ 5 = £120
So Jane receives £120 × 3 = £360.
(b) Alison receives £120 × 2 = £240.

Remember to check that the two answers (£360 and £240) add up to £600.

5 *Divide £48 into 8 equal parts.*
(a) 1 part = £48 ÷ 8 = £6
So charity A received £6 × 5 = £30.
(b) Charity B received £21 which is 3 parts, so 1 part is £7. The full amount is 8 parts which is £7 × 8 = £56.

6 (a) 60 litres of grapefruit juice
(b) 1 part = 750 ml ÷ 5 = 150 ml
So the amount of pineapple juice = 150 ml × 2 = 300 ml

7 (a) £1520 ÷ 20 = £76
John's share = £76 × 11 = £836
(b) 9 parts = £450
1 part = £450 ÷ 9 = £50
So John's share = £50 × 11 = £550

8 *Remember that 1 litre = 1000 ml.*
(a) *You need to find how many times 200 'goes into' 1000.*
1000 ÷ 200 = 5
So 5 cans are needed.
(b) (i) Cost = £3·30 × 5
= £16·50
(ii) 4 bottles are equivalent to 1 litre (1000 ÷ 250 = 4).
Cost = £3·99 × 4
= £15·96
(c) The bottle is the better value for money.

9 (a) *Large size*
500 g costs £1·35.
100 g costs £1·35 ÷ 5 = £0·27

Family size
750 g costs £1·89.
100 g costs
£1·89 ÷ 7·5 = £0·252
= £0·25 (to the nearest penny)

(b) The Family size packet

(c) *Here are three reasons. You may have an equally valid one which would gain the marks.*

A person living alone wouldn't want a Family size packet because the cereal would be stale before the packet was finished.

The Family size was too tall to fit in the cupboard.

He or she couldn't afford the Family size packet.

More help or practice

Solving ratio problems ► Book G4 pages 16 to 17, Book G7 pages 22 to 28, Book B5 pages 52 to 54, Book RB+ pages 53 to 58

Best value for money ► Book G4 pages 18 to 19, Book G7 pages 1 to 6, Book B5 pages 12 to 17

Sharing in a given ratio ► Book B5 pages 54 to 55

Mixed number (page 25)

1 12% of £3·00 = £3 × 0·12
= £0·36 or 36p

So 40p an hour extra is the better deal.

2 The fraction of rooms occupied on Wednesday is $\frac{69}{85}$.

$\frac{69}{85}$ = 0·8117 …
= 81·17 … %
= 81·2% (to 1 d.p.)

3 (a) £4·20 × 38 = £159·60

(b) $\frac{1}{4}$ of £4·20 = £4·20 ÷ 4
= £1·05

So Mary's overtime rate is £4·20 + £1·05 = £5·25 per hour.

(c) 5% of £159·60 = £159·60 × 0·05
= £7·98

Mary's new pay per week is £159·60 + £7·98 = £167·58.

4 (a) 89343 – 88156 = 1187

Cost of units used = 1187 × 8p
= 9496p
= £94·96

Total cost = £11·05 + £94·96
= £106.01

(b) 5% of £106·01 = £106·01 × 0·05
= £5·30 (to the nearest penny)

Total electricity bill = £106·01 + £5·30
= £111·31

5 (a) 146

(b) (i) 8764

(ii) 9000 (to the nearest thousand)

6 (a) £300 × 20 = £6000

(b) £307 × 19 = £5833

(c) 307 ÷ 48 = 6 rem 19

So 7 coaches will be needed.

You need to round up.

*Remember to show **all** your working as no calculator is allowed.*

7 (a) $\frac{5}{15} = \frac{1}{3}$ of the diagram is shaded.

(b) $\frac{1}{3}$ = 0·333 …
= 0·333 … × 100%
= 33·3 … %

(c) $\frac{2}{5}$ of the squares must be shaded.

$\frac{1}{5}$ of 15 = 3, so $\frac{2}{5}$ of 15 = 6.

So Stuart must shade 1 more square.

8 (a) 36% + 49% = 85%

(b) The number of colour television licences increases as the number of black and white licences decreases.

(c) 3% = $\frac{3}{100}$ did not have a licence.

9 (a) 2, 3 and 5 are prime numbers.

(b) (i) 7 tens (or 70)

(ii) 7 hundreds (or 700)

(iii) 7 units (or 7)

(c) 6

10 (a) 12649

(b) 12550

(c) $\frac{1}{3}$ of 12600 = 12600 ÷ 3
= 4200

Approximately 4200 spectators are women.

11

INPUT	OUTPUT
22	5
3	**0·25 or $\frac{1}{4}$**
$^-2$	**$^-1$**
42	10

*If you know the output and want to find the input, you need to use **inverse** operations (reverse the operations):*

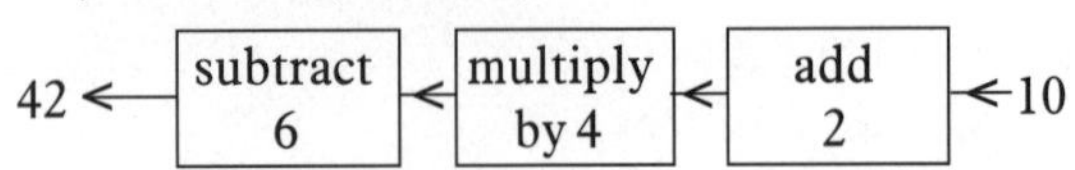

12

O.K. PLUMBERS	
Fixed call-out fee	£21·50
$1\frac{3}{4}$ hours at £16 per hour	**£28·00**
Parts	£29·75
Total before VAT	**£79·25**
VAT at $17\frac{1}{2}$%	**£13·87**
Total due	**£93·12**

13 (a) $182 \div 48 = 3{\cdot}971 \ldots$
So David will need to hire 4 coaches.

(b) (i) [1][3][×][1][8][2][=] or [1][8][2][×][1][3][=]

(ii) £2366

*You will have your own method of multiplying 182 by 13. It is very important to show **all** your working. You will not get any marks for just the answer even if it is right!*

(c) Discount = £13 × 0·15
= £1·95
Cost of entry = £13 – £1·95
= £11·05

14 (a) $\frac{1}{5}$ of £7500 = £7500 ÷ 5 = £1500

$\frac{2}{5}$ of £7500 = £1500 × 2 = £3000

*Paul paid £3000 when he bought the car. This is called the **deposit**.*

(b) Paul still owed £7500 – £3000 = £4500.
This is the amount of money he needs to 'pay off' over 12 months. To find his monthly payment divide £4500 by 12.
£4500 ÷ 12 = £375

ALGEBRA

Writing algebra (page 28)

1 (a) $3a$ (b) $2b$ (c) $6d-4$ (d) $a+3$
(e) $3m+2$ (f) $2x+7$ (g) $4c+7d$ (h) $q+4r$
(i) $6a$ (j) $30c$ (k) $4c^2$ (l) $2cd$
(m) a^3 (n) $6ab$ (o) $5xy$ (p) $2x^3$

In (o) notice that we write the number first ($5xy$, not $xy5$).

2 (a) $3x + x = 4x$
(b) $z - w$

3 (a) $x + x + x = 3x$
(b) $x + x + y + y + y = 2x + 3y$

4 (a) $c + d$ coins
(b) $m - n$ pounds or £$(m-n)$

5 Ben earns £a in 1 hour.
So in 7 hours he earns £$a \times 7$ = £$7a$

6 (a) $5 \times 2a = 10a$
(b) $5b \times 3 = 15b$
(c) $2a \times 2a = 4a^2$

7 (a) $4c + 24$
(b) $3d - 12$
(c) $4a + 2 + 3 = 4a + 5$
(d) $30 - 5x - 10 = 20 - 5x$
(e) $a + 3a - 12 = 4a - 12$
(f) $5x + 5 - 3 = 5x + 2$
(g) $12 + 8x - x = 12 + 7x$ or $7x + 12$
(h) $6a + 8b + 1$

8 (a) For 1 hour Marcus is paid £x.
So for 30 hours he is paid £$x \times 30$ = £$30x$.
(b) For 1 hour Marcus is paid £$x \times 2$ = £$2x$.
For 8 hours he is paid £$2x \times 8$ = £$16x$.
(c) His total pay in pounds $= 30x + 16x$
$= 46x$
(d) For 1 hour Christopher is paid £$(x+2)$.
For 6 hours he is paid £$6(x+2)$ or £$(6x+12)$.

More help or practice
Using shorthand ► Book G1 pages 26 to 29;
Book G9 pages 1 to 4, 115 to 116; Book B3 page 26

Coordinates and graphs (page 30)

Most of the graphs in these answers have been reduced.

1 (a) ($^-2$, 3)
(b) and (c) See the graph.
(d) (i) See the graph. (ii) ($^-3$, 1)

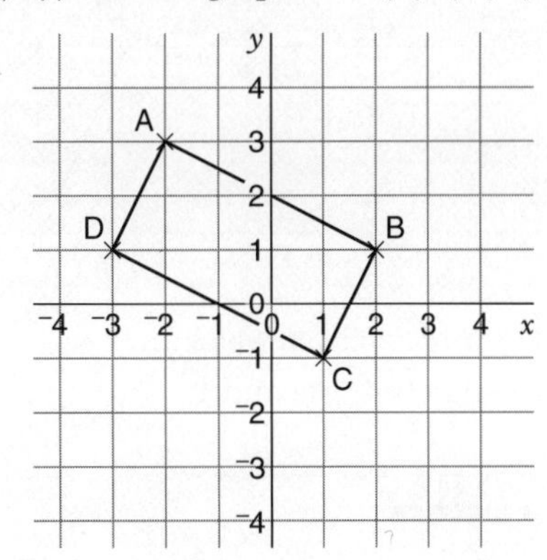

2 (a) See the graph.
(b) (i) ($^-1$, 0)
(ii) ($^-\frac{1}{2}$, $\frac{1}{2}$) or ($^-0{\cdot}5$, $0{\cdot}5$)
You can find the middle point of PQ 'by eye' or by using a ruler.
(c) See the graph.

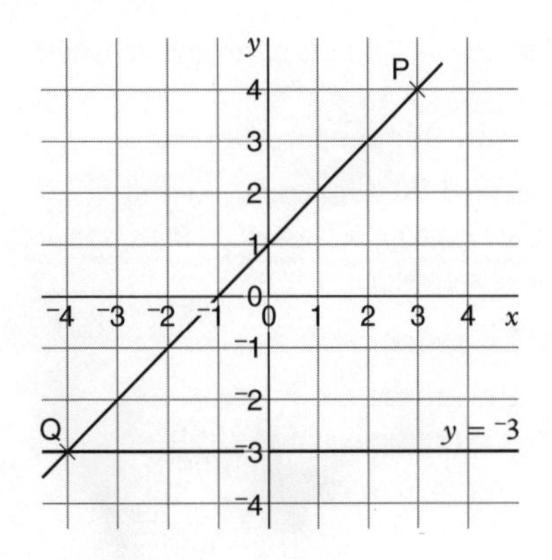

3 (a)

x	0	1	2	3	4	5
y	**3**	7	**11**	15	**19**	**23**

(b) and (c)

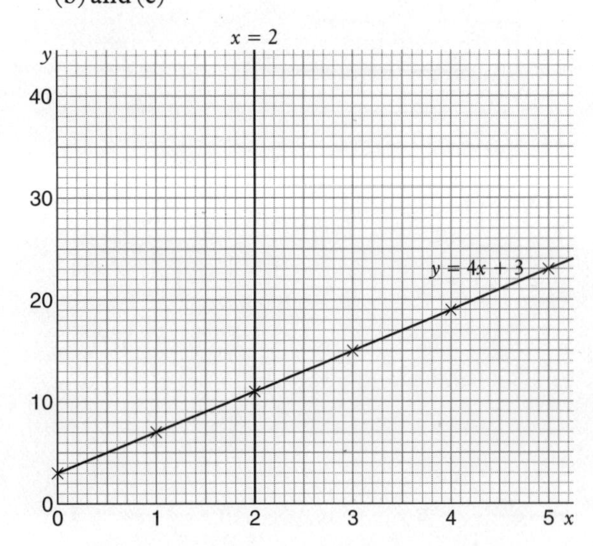

Notice that the graph of $y = 4x + 3$ is a straight line.
(d) The two lines cross at (2, 11).

4 (a) and (b)
To draw a curve you need to plot several points, but to draw a straight line you only need three points (including one as a check).
We have chosen to plot (0, 0), (2, 10) and (4, 20).
(c) (0, 0) and ($2\frac{1}{2}$, $12\frac{1}{2}$)

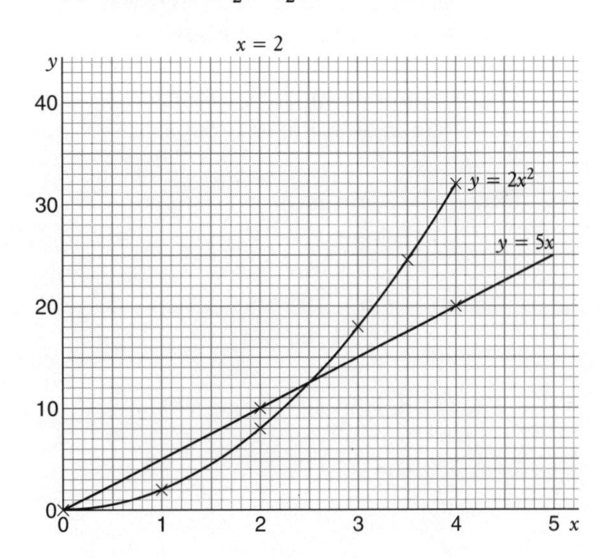

5 (a)

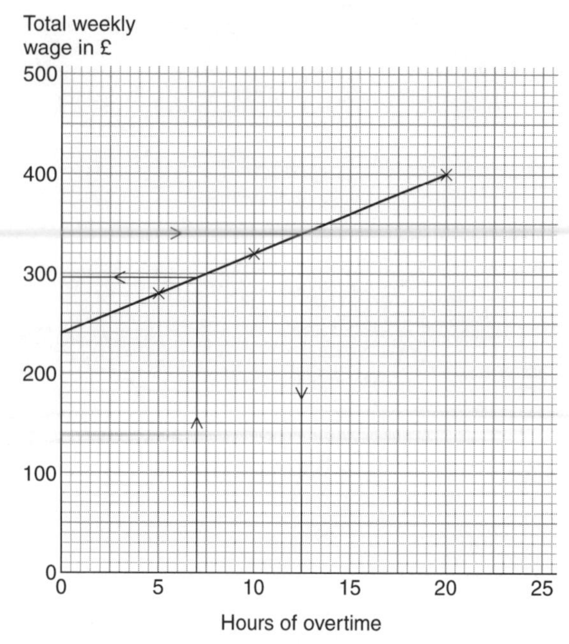

(b) (i) £296

Any answer between £290 and £300 would gain full marks.

(ii) 12·5 hours

The arrowed lines show how to take the readings.

(iii) £240

The basic wage is given by the point where the line cuts the 'up' axis. This is where the number of hours of overtime is zero.

6 (a) *To find the height when $t = 2$, substitute this value for t in the equation $h = 100 - 5t^2$; do the same for $t = 3$.*

$$\text{When } t = 2, h = 100 - (5 \times 2 \times 2) = 100 - 20 = 80$$

$$\text{When } t = 3, h = 100 - (5 \times 3 \times 3) = 100 - 45 = 55$$

The two heights are 80 m and 55 m.

(b)

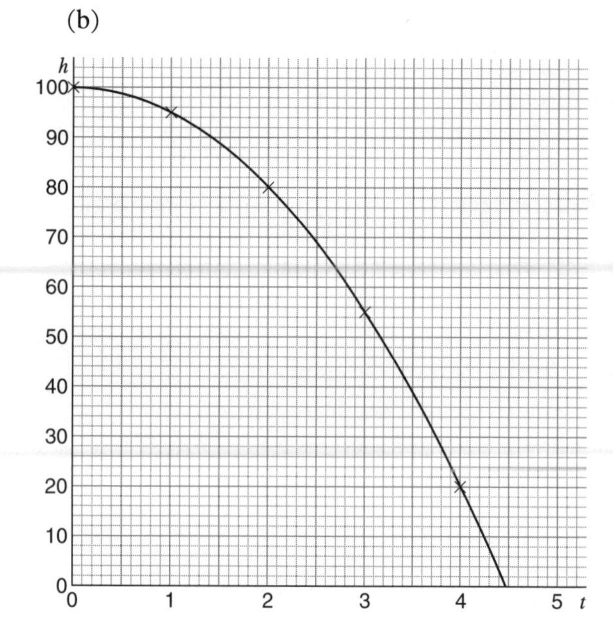

(c) The stone reaches the water after approximately $4\frac{1}{2}$ seconds.

More help or practice
Positive coordinates ► Book G1 page 49
Negative coordinates ► Book G9 pages 8 to 12
Drawing straight line graphs ► Book B1 pages 98 to 101
Drawing non-linear graphs ► Book B5 pages 71, 99 to 101

Time and conversion graphs (page 32)

All the graphs in these answers have been reduced.

1 (a) The train stopped for 15 minutes.

(b) Between 3:00 p.m. and 3:30 p.m.

The train is going fastest when the slope of the graph is steepest.

(c) (i) $2\frac{1}{2}$ hours

(ii) 100 miles

(iii) Average speed $= \dfrac{\text{distance}}{\text{time}} = \dfrac{100 \text{ miles}}{2{\cdot}5 \text{ hours}} = 40$ m.p.h.

2 (a) D

(b) The cold water tap was turned on.

(c) Both taps were turned off.

3 (a) 64 dollars (b) £15

The arrowed lines on the graph show how to take the readings.
Always take care with the scales on the axes.
For this graph each small square on the £ axis stands for £1 and each small square on the $ axis stands for $2.

4 (a) (i) 32°C (ii) 27°C

Each small square on the temperature axis represents 1 °C.

(b) $11\frac{1}{2}$ minutes

Each small square on the time axis represents $\frac{1}{2}$ minute.

(c) After $16\frac{1}{2}$ minutes

(d) 5 minutes

This is the difference between your answers to parts (b) and (c).

5 (a)

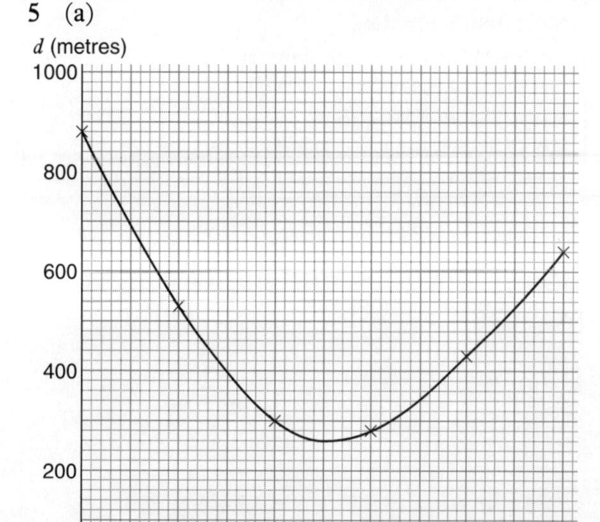

(b) (i) Approximately 550 m

Any answer between 540 m and 560 m would gain full marks.

(ii) The boats are 400 m or less apart between 1·5 minutes and 3·75 minutes, that is for 2·25 minutes or $2\frac{1}{4}$ minutes.

(iii) The boats are closest to one another when $t = 2{\cdot}5$ minutes and $d = 250$ metres.

6 (a) 39% of men smoked in 1976.

(b) Between 1976 and 1980

The slope of the graph is steepest between these years.

(c) Between 1988 and 1992

The graph slopes up.

(d)

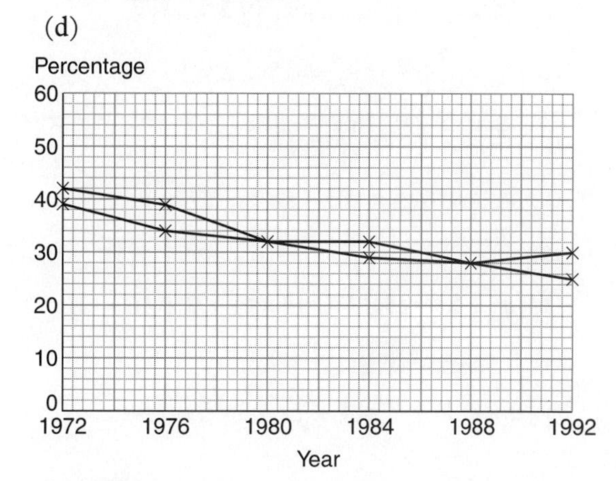

(e) 1984

More help or practice

Conversion graphs ► Book G7 pages 34 to 35
Time graphs ► Book G7 page 40 to 42
Interpreting and drawing graphs
► Book B2 pages 45 to 52, Book B3 pages 14 to 22

Using formulas and expressions (page 35)

1 (a) $2b = 2 \times 2 = 4$
(b) $a + b = 5 + 2 = 7$
(c) $a - c = 5 - 3 = 2$
(d) $3a + 5 = (3 \times 5) + 5 = 15 + 5 = 20$
(e) $b^2 = b \times b = 2 \times 2 = 4$
(f) $4c - 6 = (4 \times 3) - 6 = 12 - 6 = 6$
(g) $4b + 3c = (4 \times 2) + (3 \times 3) = 8 + 9 = 17$
(h) $7c - 4a = (7 \times 3) - (4 \times 5) = 21 - 20 = 1$
(i) $2a^2 = 2 \times a \times a = 2 \times 5 \times 5 = 50$
Notice the difference between $2a^2$ and $(2a)^2$.
The latter would have value $(2 \times 5)^2 = 10^2 = 100$.
(j) $c^2 - 1 = (c \times c) - 1 = (3 \times 3) - 1 = 9 - 1 = 8$

2 (a) (i) $(3 \times 2) + 5 = 6 + 5 = 11$
(ii) $(a \times 2) + 5 = 2a + 5$
(b) $a \rightarrow \boxed{\times 2} \rightarrow \boxed{+3} \rightarrow 2a + 3$

3 (a) $\frac{c}{a} = \frac{6}{3} = 6 \div 3 = 2$
(b) $\frac{4a}{c} = \frac{4 \times 3}{6} = \frac{12}{6} = 12 \div 6 = 2$
(c) $ab + c = (3 \times 4) + 6 = 12 + 6 = 18$
(d) $\frac{ac}{2} = \frac{3 \times 6}{2} = \frac{18}{2} = 18 \div 2 = 9$
(e) $\frac{3b}{a} = \frac{3 \times 4}{3} = \frac{12}{3} = 12 \div 3 = 4$
(f) $a^3 = a \times a \times a = 3 \times 3 \times 3 = 27$
(g) $\frac{2b - a}{5} = \frac{(2 \times 4) - 3}{5} = \frac{8 - 3}{5} = \frac{5}{5} = 1$
(h) $a^2 + b^2 = (3 \times 3) + (4 \times 4) = 9 + 16 = 25$
(i) $b + \frac{c}{a} = 4 + \frac{6}{3} = 4 + 2 = 6$
(j) $a(b + 1) = 3 \times (4 + 1) = 3 \times 5 = 15$

4 *Substitute $r = 40$ into the formula $A = 3 \times r \times r$.*
$A = 3 \times 40 \times 40 = 4800$
Approximate area = 4800 cm^2

5 (a) When $C = 18, F = (2 \times 18) + 30 = 36 + 30 = 66$
(b) When $C = 0, F = (2 \times 0) + 30 = 30$
Remember that multiplying by zero always gives zero, whereas adding zero leaves a number unchanged.

6 $t = 20w + 30$

7 time $= \frac{12 \times 30}{20} = \frac{360}{20} = 360 \div 20 = 18$
She can stay under water 18 minutes.

8 $k = 70 \times 1{\cdot}6 = 112$
70 miles is roughly equal to 112 km.

9 $t = \frac{d}{s} = \frac{330}{55} = 330 \div 55 = 6$
Daniel takes 6 hours.

10 (a) The formula is $A = (h \times w) \times 0{\cdot}8$ or
$A = hw \times 0{\cdot}8$ or
$A = 0{\cdot}8hw$
(b) *Substitute $w = 10$ and $h = 15$ in the formula.*
$A = 0{\cdot}8 \times 10 \times 15 = 120$
An approximate area is 120 cm^2.

11 *Substitute $p = 15$ into the formula.*
$N = \frac{p(p-1)}{2} = \frac{15 \times 14}{2} = 105$

12 $V = Ah = 24{\cdot}4 \times 10{\cdot}5 = 256{\cdot}2$
Volume = 256·2 cm^3
cm^3 stands for cubic centimetres.

More help or practice

Substitution ► Book G9 pages 102 to 107;
Book B1 pages 94 to 97, 102 to 104; Book B2 pages 71 to 73;
Book B3 pages 26 to 34; Book B4 pages 5 to 6;
Book B5 pages 96 to 98
Forming and manipulating expressions
► Book G9 pages 5 to 7, Book B+ pages 73 to 81

Solving linear equations (page 38)

1 (a) $4x = 32$
$x = 8$ *(dividing both sides by 4)*

(b) $x + 7 = 15$
$x = 8$ *(subtracting 7 from both sides)*

(c) $x - 8 = 20$
$x = 28$ *(adding 8 to both sides)*

(d) $2x + 1 = 9$
$2x = 8$ *(subtracting 1 from both sides)*
$x = 4$ *(dividing both sides by 2)*

(e) $3x - 4 = 11$
$3x = 15$ *(adding 4 to both sides)*
$x = 5$ *(dividing both sides by 3)*

(f) $7x - 14 = 0$
$7x = 14$ *(adding 14 to both sides)*
$x = 2$ *(dividing both sides by 7)*

(g) $22 = 5x + 2$
$5x + 2 = 22$ *(change the equation round)*
$5x = 20$ *(subtracting 2 from both sides)*
$x = 4$ *(dividing both sides by 5)*

(h) *Start by multiplying out the brackets.*
$2(x - 3) = 1$
$2x - 6 = 1$
$2x = 7$ *(adding 6 to both sides)*
$x = 3\frac{1}{2}$ or $3{\cdot}5$ *(dividing both sides by 2)*

(i) $5(2x + 1) = 30$
$10x + 5 = 30$ *(multiplying out the brackets)*
$10x = 25$ *(subtracting 5 from both sides)*
$x = 25 \div 10$ *(dividing both sides by 10)*
$= 2{\cdot}5$ or $2\frac{1}{2}$

2 (a) $2x + 1 = x + 9$
$x + 1 = 9$ *(subtracting x from both sides)*
$x = 8$ *(subtracting 1 from both sides)*

(b) $3x - 7 = x + 5$
$2x - 7 = 5$ *(subtracting x from both sides)*
$2x = 12$ *(adding 7 to both sides)*
$x = 6$ *(dividing both sides by 2)*

(c) $5x - 9 = 2x + 6$
$3x - 9 = 6$ *(subtracting 2x from both sides)*
$3x = 15$ *(adding 9 to both sides)*
$x = 5$ *(dividing both sides by 3)*

(d) $2 + 5x = 10 - 3x$
$2 + 8x = 10$ *(adding 3x to both sides)*
$8x = 8$ *(subtracting 2 from both sides)*
$x = 1$ *(dividing both sides by 8)*

(e) $10 + 2x = x + 14$
$10 + x = 14$ *(subtracting x from both sides)*
$x = 4$ *(subtracting 10 from both sides)*

(f) $2(x + 1) = x + 5$
$2x + 2 = x + 5$ *(multiplying out the brackets)*
$x + 2 = 5$ *(subtracting x from both sides)*
$x = 3$ *(subtracting 2 from both sides)*

3 $4n - 3 = 13$
$4n = 16$ *(adding 3 to both sides)*
$n = 4$ *(dividing both sides by 4)*

4 (a) $a + (a + 3) + (2a - 1) = 22$
$a + a + 3 + 2a - 1 = 22$
$4a + 2 = 22$
$4a = 20$
$a = 5$

(b) 8 cm and 9 cm
Substitute $a = 5$ into $a + 3$ and $2a - 1$ to find the lengths of the other two sides.

5 Let the number be n.
Then $3n + 5 = 23$
$3n = 18$
$n = 6$

6 (a) $C = 20n + 30$
Substitute $n = 15$ in this equation.
$C = (20 \times 15) + 30$
$= 300 + 30$
$= 330$
The cost is 330p or £3·30.

(b) (i) $250 = 20n + 30$
(ii) *To solve the equation it is probably best to write the equation so that the expression in n is on the left-hand side.*
$20n + 30 = 250$
$20n = 220$
$n = 11$
The number of words is 11.

7 *The angles of a triangle add up to 180°.*

(a) $x + 2x + 3x = 180°$
$6x = 180°$
$x = 30°$

(b) The angles of the triangle are 30°, 60° and 90°.

8 Let the number of cars in the car park be n.
Then $n + 55 = 6n$ or $6n = n + 55$.
Solve $6n = n + 55$.
$5n = 55$
$n = 11$

More help or practice
Balance problems ▸ Book B2 pages 53 to 58
Solving equations ▸ Book B+ pages 68 to 72

Sequences and terms (page 40)

1 (a)

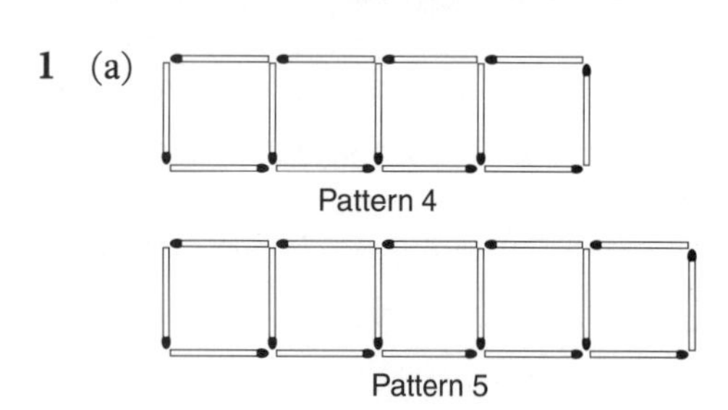

(b)

Pattern	1	2	3	4	5
Number of rods	4	7	**10**	**13**	**16**

(c) The 8th pattern can be made with 25 matches.
You can work this out by counting on in threes until you get to 25.

(d) The number of matches is always equal to '3 times the pattern plus 1'.
So pattern 16 = $(3 \times 16) + 1 = 49$.
You would also get full marks for stating that pattern 16 = 4 + (15 × 3) = 4 + 45 = 49.

2 (a) The next three terms are 16, 25 and 36.

(b) The numbers are square numbers.
$1^2 = 1$, $2^2 = 4$, $3^2 = 9$, $4^2 = 16$, $5^2 = 25$, $6^2 = 36$.

3 (a) The 5th stack will have (10 + 5) bricks = 15 bricks.

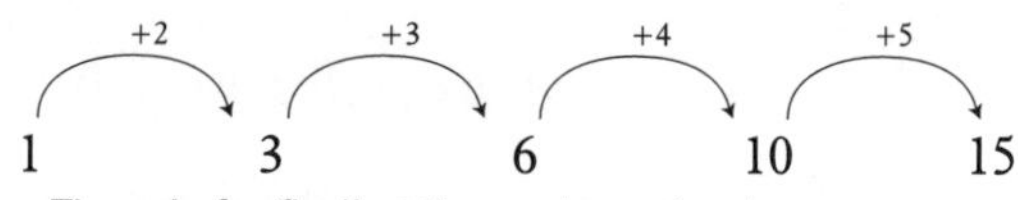

The rule for finding the next number is 'Add on one more than last time'.

(b) The 6th stack has 21 bricks so the 7th stack has 28 bricks.

4 (a) (i) 21, 26 (ii) 32, 64 (iii) 15, 12

(b) (i) Add 5 to the number before.
(ii) Multiply the number before by 2.
(iii) Subtract 3 from the number before.

5 (a) 30, 36

(b) Each term is 6 times its position number.
If you find it difficult to work out the rule, a table like this might help.

Position number	*1*	*2*	*3*	*4*	*5*	*6*
Term	*6*	*12*	*18*	*24*	*30*	*36*

The sequence is the '6 times table'.

(c) (i)

Position number	*1*	*2*	*3*	*4*	*...*	*20*
Term	*4*	*8*	*12*	*16*	*...*	*80*

Each term is 4 times its position number.
The 20th term is $20 \times 4 = 80$.

(ii) The nth term is $n \times 4 = 4n$.

6 (a) (i) $(38 + 1) \times 2 = 39 \times 2 = 78$
You could have answered part (i) by using a flow diagram.

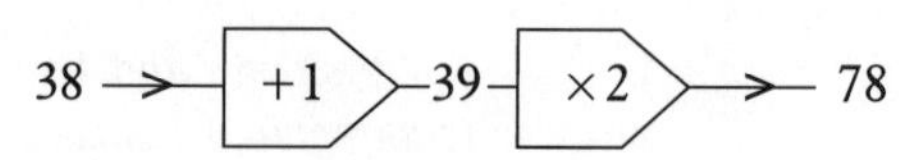

(ii) *To find the starting number you need to reverse the flow diagram and use the inverse operations, divide by 2 and subtract 1.*

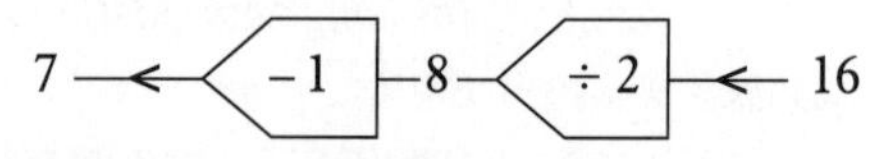

The starting number is 7.

(b) (i) Add 4 to the number before.
(ii) $4n + 1$
Each term is 4 times its position number plus 1.

Position number	*1*	*2*	*3*	*4*	*...*	*n*
Term	*5*	*9*	*13*	*17*	*...*	*4n + 1*

7 (a) (i) Add 3 (ii) Add 2 (iii) Add 5

(b) (i) $3n$ (ii) $2n + 1$ (iii) $5n - 1$

Notice the connection between your answers to (a) and your answers to (b).

> **More help or practice**
> Continuing sequences and describing rules
> ► Book G9 pages 109 to 111, Book G+ pages 33 to 37, Book RB+ pages 12 to 16

Mixed algebra (page 42)

1 (a) *One way to answer this is to form an equation.*

Let the unknown number be a.
Then $5a + 4 = 29$
$5a = 25$
$a = 5$

Another way is to use flow diagrams.

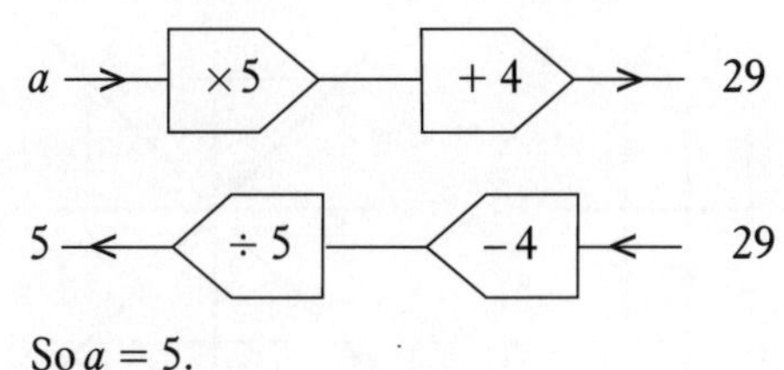

So $a = 5$.

(b) $(2 \times {}^{-}1) + (3 \times 4) = {}^{-}2 + 12 = 10$

You can think of $2 \times {}^{-}1$ as ${}^{-}1 + {}^{-}1 = {}^{-}2$.

(c) (i) $2a + 3b - 4a = 3b - 2a$

(ii) $3x + 2(2x - 3y) = 3x + 4x - 6y$
$= 7x - 6y$

2 (a) *Substitute $m = 160$ in the equation $C = 100 + 2m$.*

$C = 100 + (2 \times 160)$
$= 100 + 320$
$= 420$

The cost is £420.

(b) Total paid $= £10 \times 35 + £5 \times 10$
$= £350 + £50$
$= £400$

(c) (i) The club made a loss.

(ii) The loss $= £420 - £400 = £20$

3 $p = 2(a + b)$
$= 2(4{\cdot}5 + 4{\cdot}2)$
$= 2 \times 8{\cdot}7$
$= 17{\cdot}4$

The perimeter is 17·4 cm.

4 (a) $L = 0{\cdot}02S^2$
$= 0{\cdot}02 \times 100 \times 100$
$= 200$

The length of the skid is 200 m.

(b)

Speed (S m.p.h.)	S^2	$0{\cdot}02S^2$	Too small	Too big
600	360000	7200	✓	
700	490000	9800		✓
680	462400	9248	✓	
690	476100	9522	✓	
692	478864	9577	✓	
693	480249	9605		✓

The speed before the skid was 693 m.p.h. (to the nearest 1 m.p.h.).

This is the speed which gives a skid length closest to 9600 m.

You may have used different numbers for your trials or may have needed to use more numbers.

5 (a)

s	0	20	40	60	80
p	25	125	225	325	425

(b)

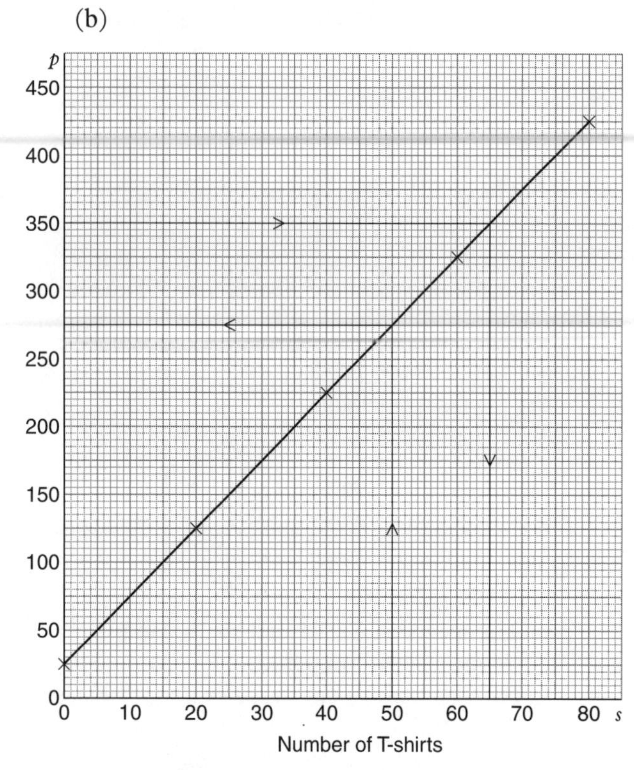

(c) (i) £275
(ii) £5·50 per T-shirt

(d) 65 T-shirts

6 (a) $1 + 3 + 5 + 7 + 9 = 25$

(b) $1 + 3 + 5 + 7 + 9 + 11 = 36$

(c) (i) odd numbers
(ii) square numbers

(d) (i) $2n - 1$
(ii) n^2

7 (a) $A = 0{\cdot}5ef = 0{\cdot}5 \times 40 \times 90$
$= 1800$
The area is $1800\,\text{cm}^2$.

(b) If the unknown length is e cm, then we know that
$0{\cdot}5 \times e \times 100 = 50e = 3000$
So $e = 3000 \div 50$
$= 60$
The other stick should be 60 cm long.

SHAPE, SPACE AND MEASURES

Understanding shape (page 44)

1 *Always try to be as accurate as you can when naming shapes.*
For example, A could correctly be called a kite, or a trapezium or just a quadrilateral.
But it is most helpful to call it a rhombus.

The answers given here are the most accurate ones.

A Rhombus
B Regular pentagon
C Trapezium
D Isosceles triangle
E Equilateral triangle
F Parallelogram
G Rectangle

With symmetry questions, it is a good idea to use tracing paper or a mirror to check your answers.

2

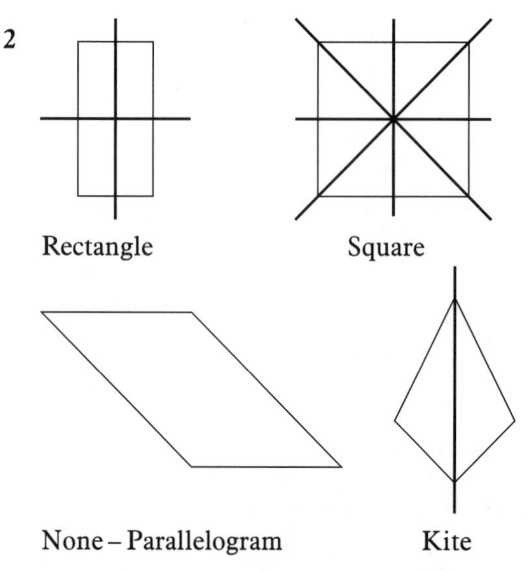

Use a mirror to check if you are not sure.

Note that you could have other names for each shape. For example, the square is also a rectangle, or even just a quadrilateral. But it is most helpful to give the most accurate name.

3 (a) Yes (b) No (c) No
(d) Yes (e) No

Check carefully here. For example, when you turn the three of spades around, the centre spade symbol turns upside down. So it does not have rotation symmetry.

4 (a) 4 planes *(like the square in question 2)*
(b) 1 plane (c) 2 planes

5 (a) 3 (b) 4 (c) 2

6

Name	Diags	Lines	Rot.
Parallelogram	**No**	**None**	2
Square	Yes	4	**4**
Rectangle	No	2	2

7

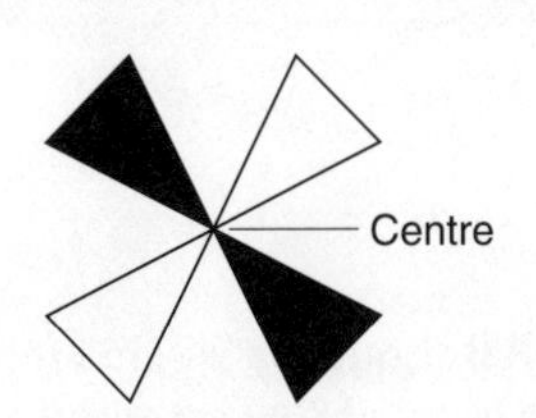

(a) 2

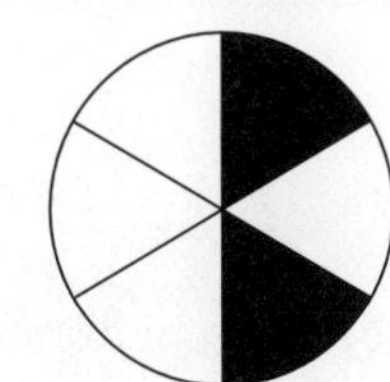

(b) None

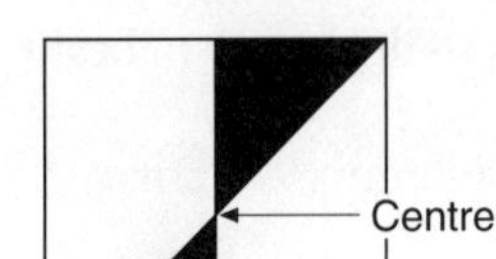

(c) 2

(d) 3

8 (a)

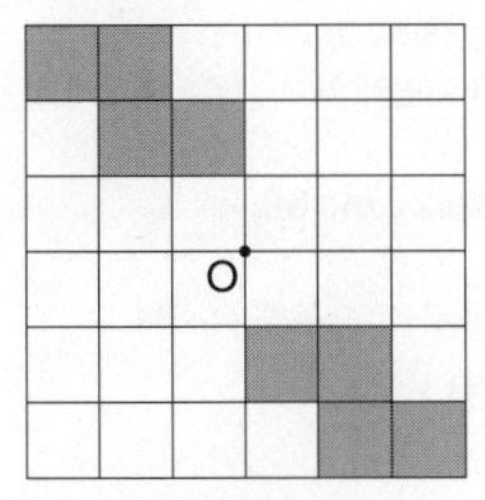

(b)

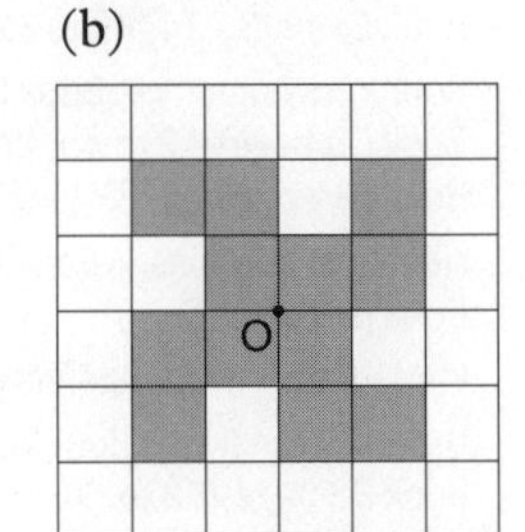

(c)

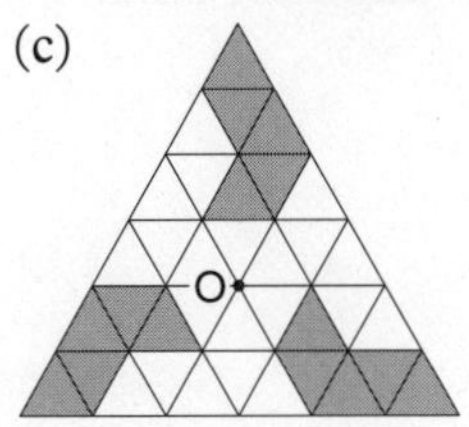

9 (a) (i)

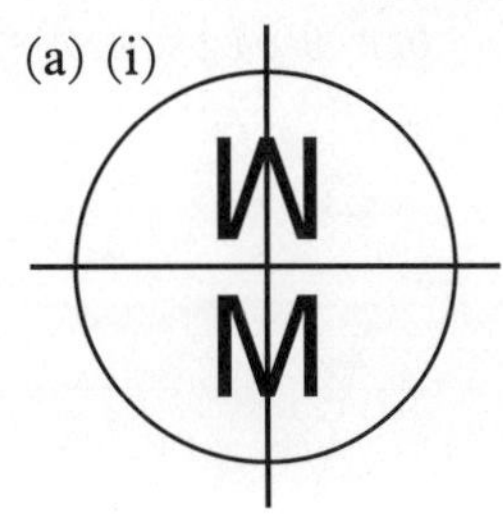

(b)

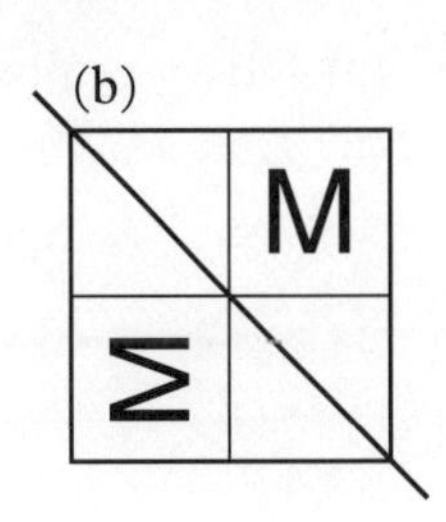

(ii) Order 2

10 Your own drawing of a regular hexagon

You could draw this in several ways. For example, you could mark every 60° round an angle measurer.
If you are asked to use only compasses and ruler, you would need to construct the hexagon like this:

Leave your construction lines showing.

Check that you can see how this diagram was constructed.

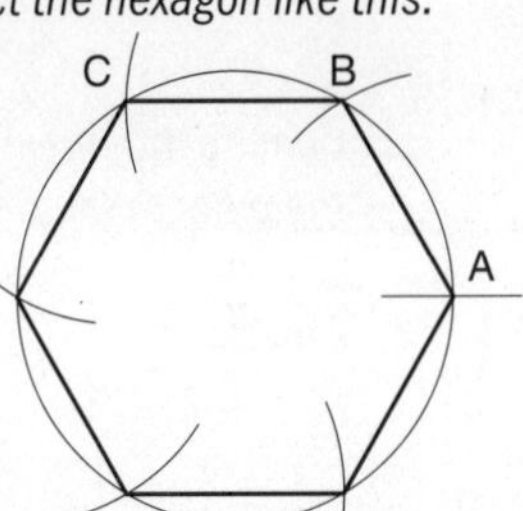

11 *Drawn half-size*

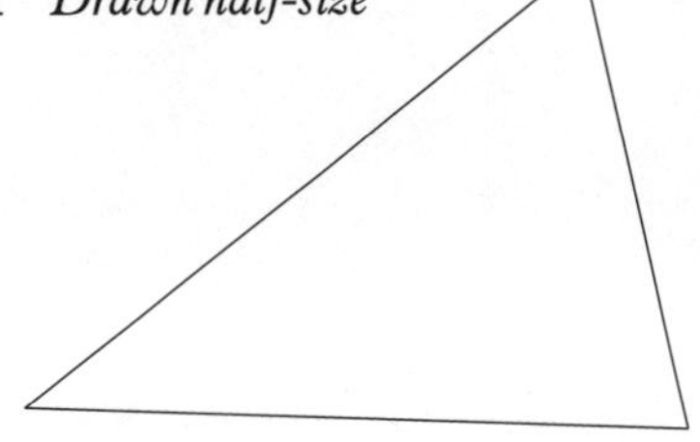

12 A drawing of an equilateral triangle

You can draw this in a similar way to question 10, but join every other point to make only three sides.

Or you can draw a line, and then use compasses to make a construction like this.

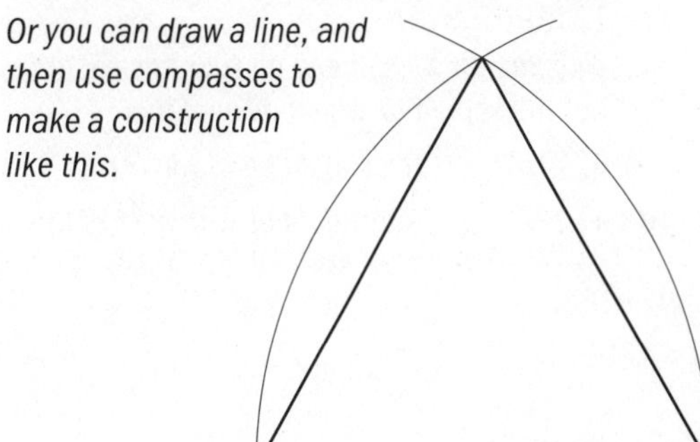

13 (a) DEF is equilateral. *(It is also isosceles.)*

(b) *Drawn half-size*

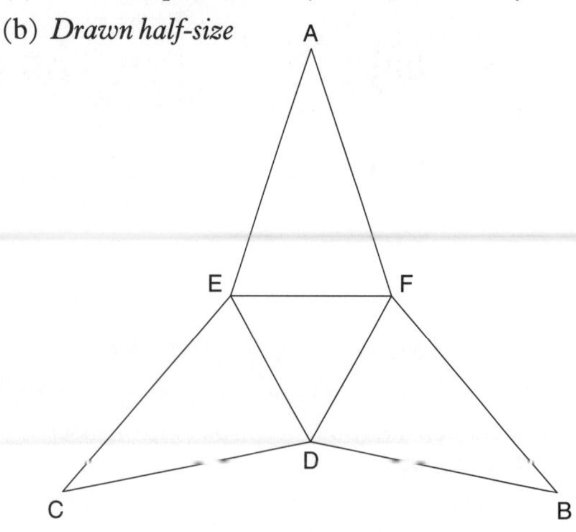

Construct triangle EFD first and then the other three triangles.

14 (a) *Drawn* ***full*** *size*

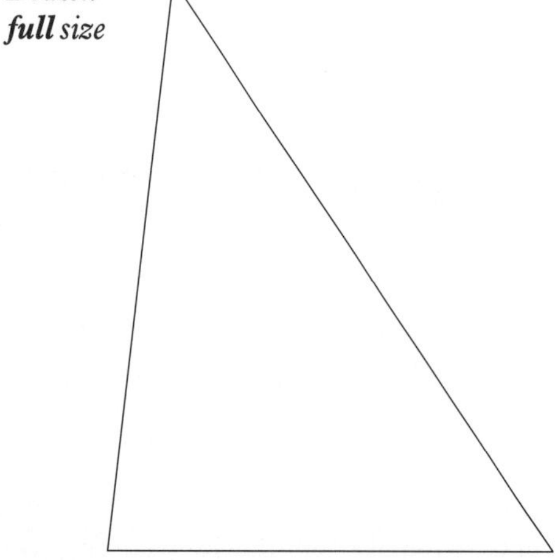

(b) 6·2cm

With care you should be within 0·2cm of this.

(c) 1240m

Your answer will depend on your answer to (b) but you should be within 40m of this.

(d) 3740m *You should be within 50m.*

(e) 41·5° *You should be within 2° of this.*

15 *Drawn half-size*

(a) *First draw BC 12cm long. Then with centre C, draw an arc, radius 9cm. Then with centre B, draw an arc, radius 12cm.*

(b) (i) 113° *You should be within 2° of this.*

(ii) 0·9m

16 (a) *Drawn half-size*

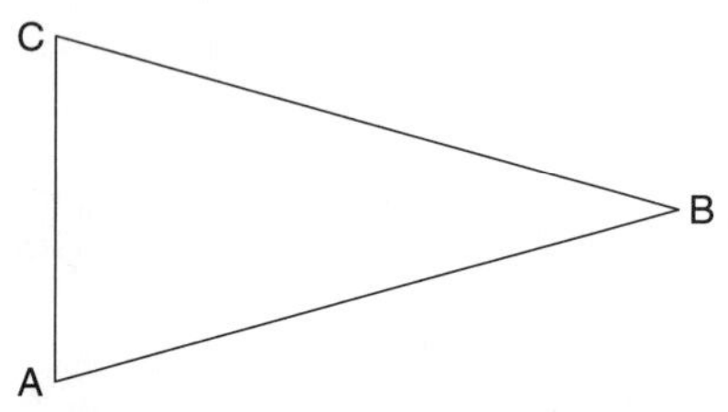

(b) On your drawing, AB should be 9·7cm long.

With care, you should be within 2mm of this.

(c) On the real pennant, AB would be $4 \times 9{\cdot}7$cm = 38·8cm, so the length of braid needed would be 20cm + 38·8cm + 38·8cm = 97·6cm.

More help or practice

Reflection symmetry ► Book G2 page 39; Book G8 pages 68 to 69; Book G9 pages 38 to 41; Book G+ pages 5 to 8; Book B1 pages 8 to 9, 74 to 77; Book B+ pages 24, 49 to 52

Rotation symmetry ► Book G8 pages 82 to 86; Book G9 pages 42 to 45; Book B1 pages 10 to 14, 78 to 82; Book B+ pages 25, 53 to 55

Names of polygons ► Book G+ pages 57 to 58, Book B+ page 15

Properties of quadrilaterals ► Book B+ pages 62 to 64

Drawing triangles ► Book G9 pages 19 to 21, Book B3 pages 63 to 64

Representing three dimensions
(page 48)

1 (a) (i) *Drawn half size*

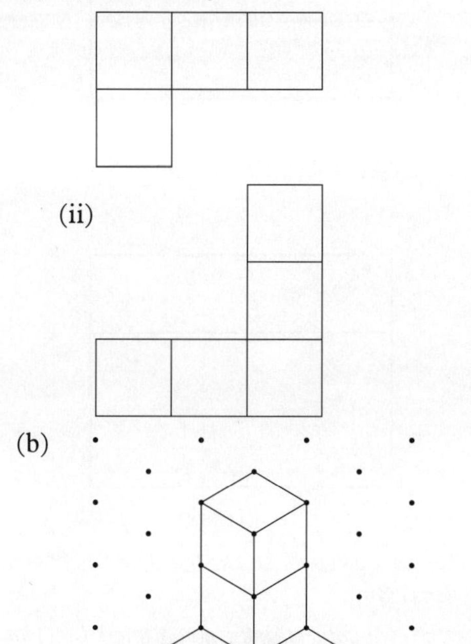

(ii)

(b)

Be careful to get your spotty paper the right way round before you begin drawing.

2

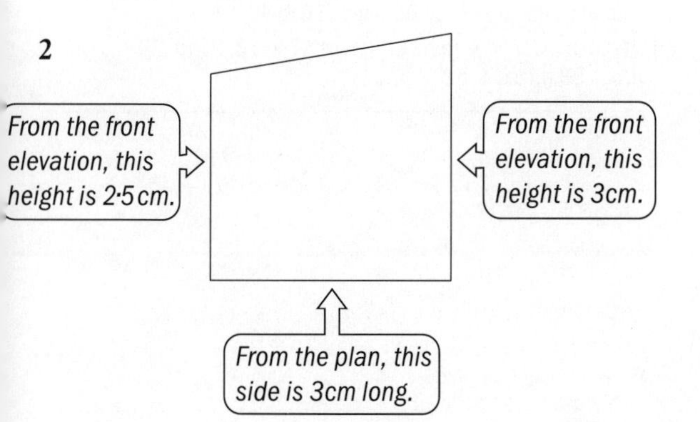

3 A and B fold up to make cubes.

These can be very hard to 'see'! If you find them difficult, you can draw the nets on squared paper, and use scissors to cut them out and make them.

4

24 × 500ml

WITH CARE

THIS SIDE UP

GLASS

OPEN OTHER END

FRAGILE

If you find this difficult, draw a rough sketch and cut it out. Then write on your sketch and try folding it up.

5 *Drawn half-size*

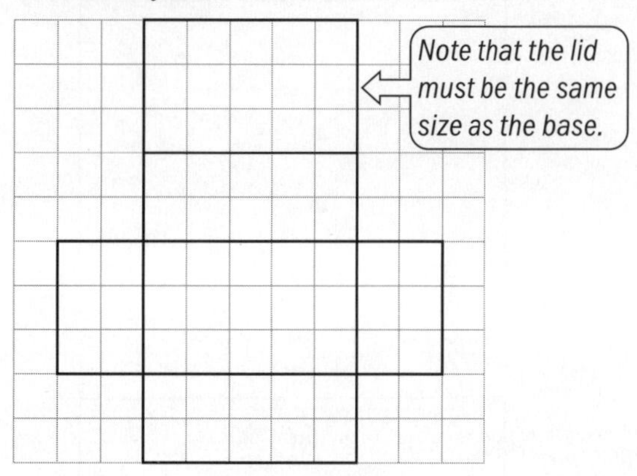

6 *Drawn half-size*

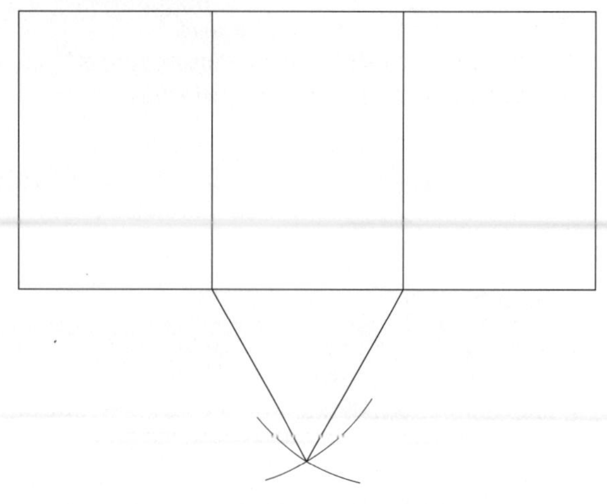

The sides of your equilateral triangle must be 5 cm long. Leave your construction lines showing.

The triangle could be attached to any of the bottom (or top) edges.

7 *Drawn half-size*

(a)

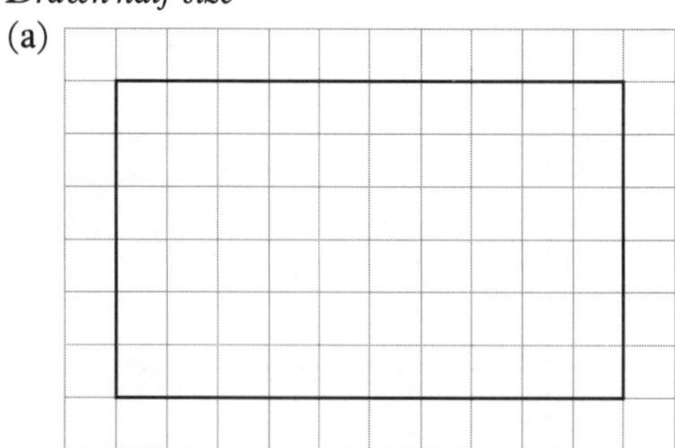

(b)

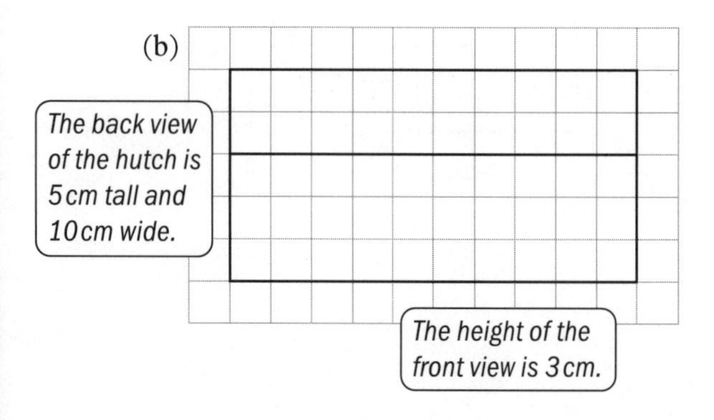

8 *Drawn half-size*

(a)

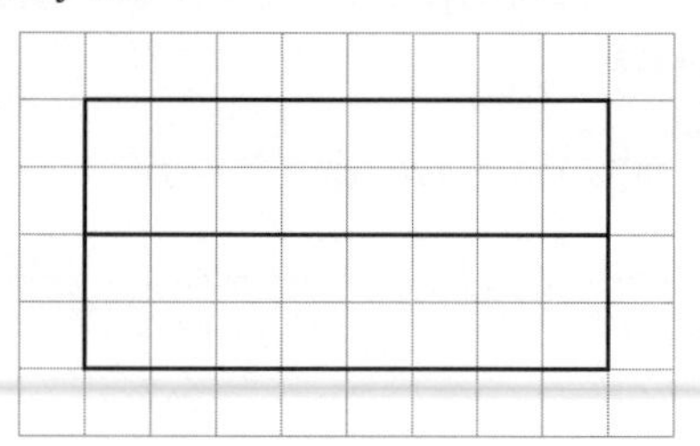

Don't forget to show the roof line.

(b)

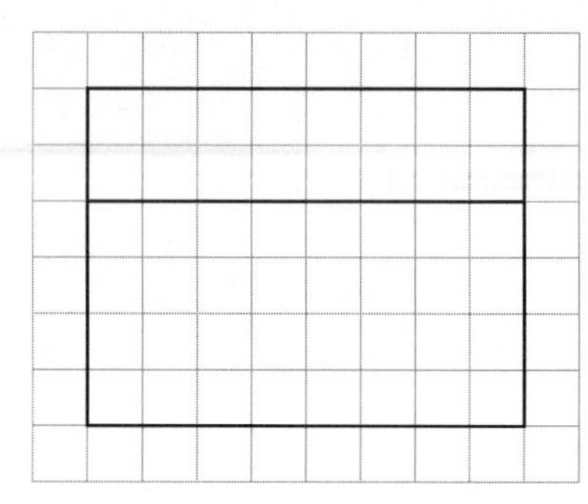

Note that you don't need to use the slant height of the roof given (2·8 m).

The sketch is not at all like the actual shed. Don't let this put you off!

More help or practice

Views ► Book G4 pages 38 to 50, Book B3 pages 8 to 13, Book B4 pages 8 to 12

Drawings and plans ► Book G5 pages 43 to 49

Plans and elevations ► Book B4 pages 78 to 87

Net of a cuboid ► Book G8 pages 36 to 40

Net of a pyramid ► Book G9 pages 17 to 18, 21 to 25; Book B3 pages 1 to 5

Angles (page 50)

1 $a = 45°$ *(180° − 75° − 60°)*
$b = 42°$ *(180° − 80° − 58°)*
$c = 139°$ *(360° − 110° − 45° − 66°)*
$d = 65°$ *(isosceles triangle)*
$e = 50°$ *(180° − 2 × 65°)*
$f = 25°$ *(90° − d)*

2 $a = 54°$ *(180 − 75° − 51°)*
$b = 126°$ *(180° − a)*
$c = 65°$ $\frac{1}{2}(180° - 50°)$
$d = 115°$ *(180° − c)*
$e = 52°$ *(180° − 128°)*

3 $q = 95°$ *(angles on a straight line)*
$r = 50°$ *(alternate or Z angles)*
$s = 135°$ *(angles on a straight line)*
$t = 85°$ *(opposite angle of 85°)*

4 (a) (i) The angles in a quadrilateral add up to 360°.
So $p = (360° - 125° - 95°) \div 2$
$= 140° \div 2$
$= 70°$
(ii) $q = 360° - 95° = 265°$
(b) (i) 100° (ii) 35° (iii) 65°

5 (a) $x = 180° - 68° = 112°$ (angles on a straight line)
(b) The two angles the legs make with the top of the board are equal, so each $= \frac{1}{2}(180° - 68°) = 56°$.
$y + 56° = 180°$ (interior angles of parallel lines), so $y = 124°$

6 (a) Triangle PED is isosceles, so
$2r + 40° = 180°$.
So $r = 70°$
(b) The exterior angle of a regular hexagon
$= 360° \div 6 = 60°$
So the interior angle $s = 180° - 60° = 120°$
(c) $\angle EDC = \angle s = 120°$
$\angle PDE = \angle r = 70°$
So $t = 120° - 70° = 50°$

7 The exterior angle of an octagon is
$360° \div 8 = 45°$,
so the interior angle $f = 180° - 45° = 135°$.
f and g add up to 180°, as they are interior angles of parallel lines.
So $g = 45°$
$\angle CQE = f = 135°$, and $\angle CQE + 90° + h = 360°$
So $h = 360° - 135° - 90°$
$= 135°$

8 (a) $A + B + C + D = 360°$, and clearly $B = D$
So $B + D = 360° - 100° - 40° = 220°$
So $B = 110°$
(b) (i) $x = 125°$ *(opposite angles of a parallelogram)*
(ii) $y = 70°$ *(one of the equal angles of the isosceles triangle STP)*
(iii) $z = 165°$ *(angles at point P, ∠SPT = y = 70°)*

More help or practice

Types of angle ► Book G+ pages 55 to 56
Basic angle properties ► Book G9 pages 59 to 60, Book B2 pages 59 to 62
Opposite, alternate and corresponding angles ► Book G9 pages 58, 60 to 63; Book RB+ pages 27 to 28
Angle properties of triangles and quadrilaterals ► Book G9 pages 80 to 84, Book B2 pages 63 to 65
Calculating angles of regular polygons ► Book G2 pages 20 to 25, Book B2 pages 66 to 68, Book B+ pages 16 to 23

Transformations (page 53)

1 (a)

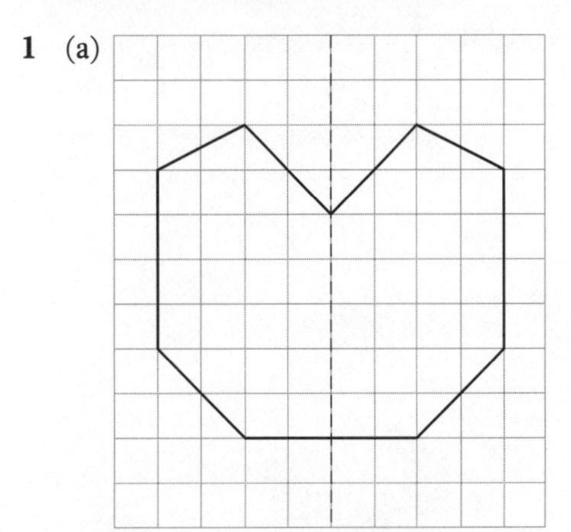

(b)

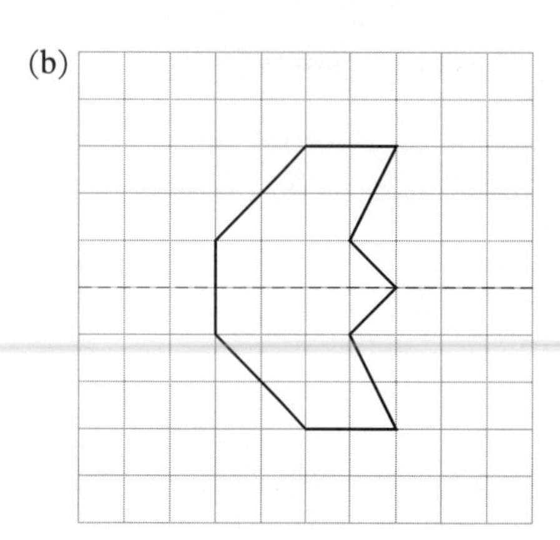

2 (a) C, D and G

(b) Any shape exactly the same size and shape as the one already on the grid

3 (a)

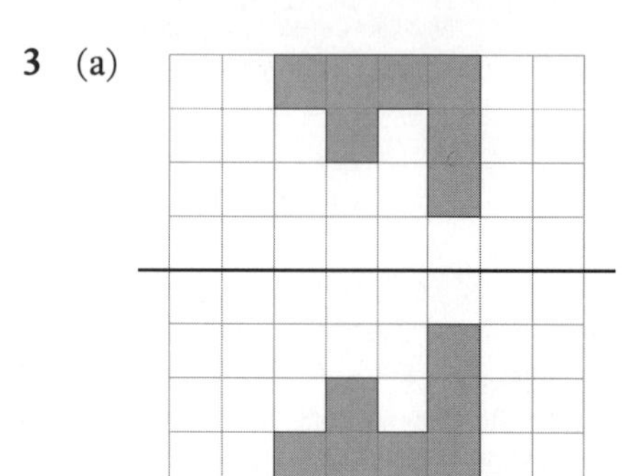

(b)

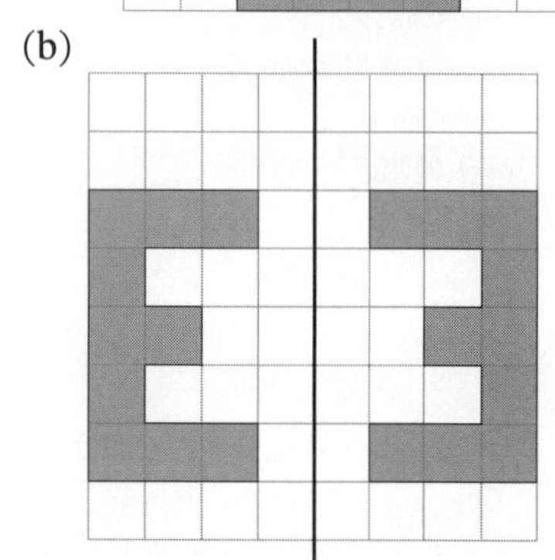

(c)

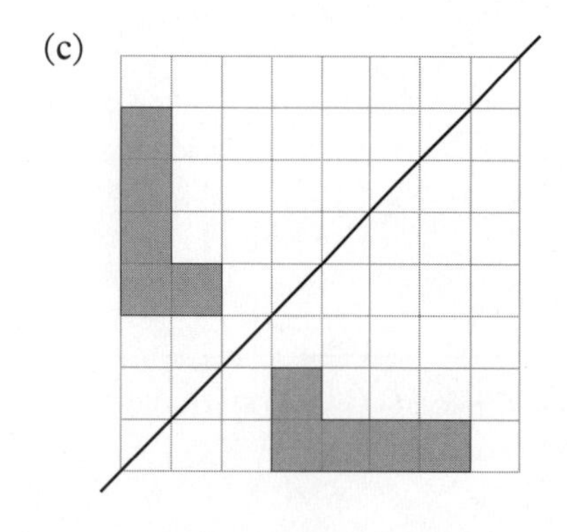

(d)

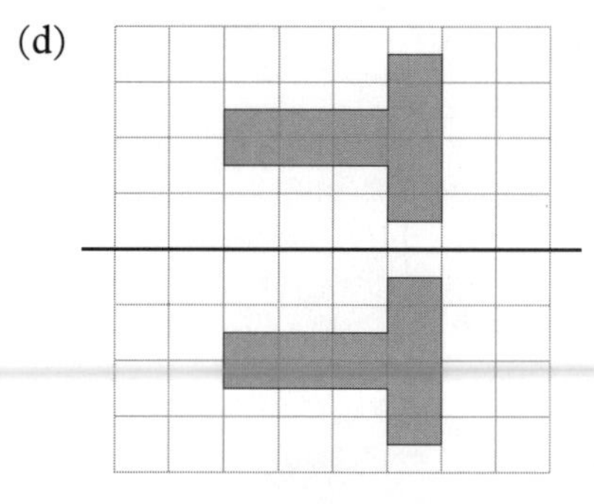

4

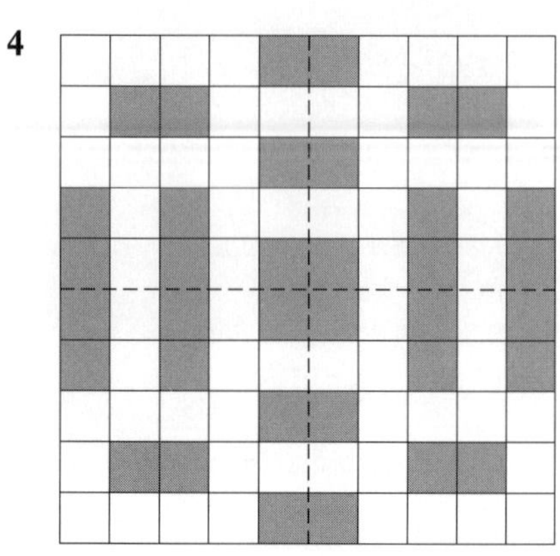

5 (a)
(b)

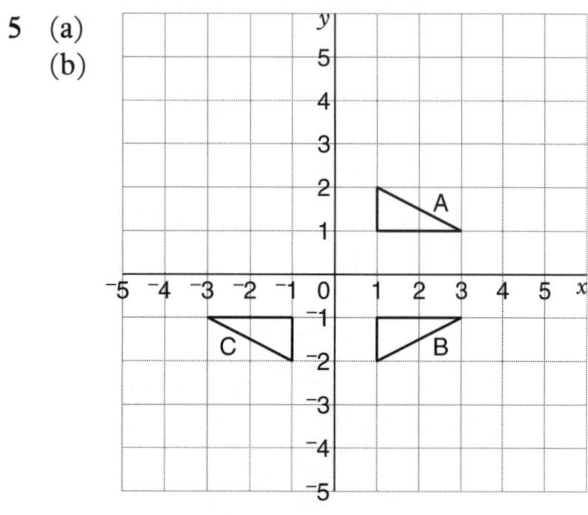

(c) A rotation about (0, 0) of 180°

Always give the centre and angle to describe a rotation. Say whether the angle is clockwise or anti-clockwise. (We don't need to say this here, because 180° is the same whichever way you go.)

6 (a)
(b)

(c) A rotation about (0, 0) of 180°

7

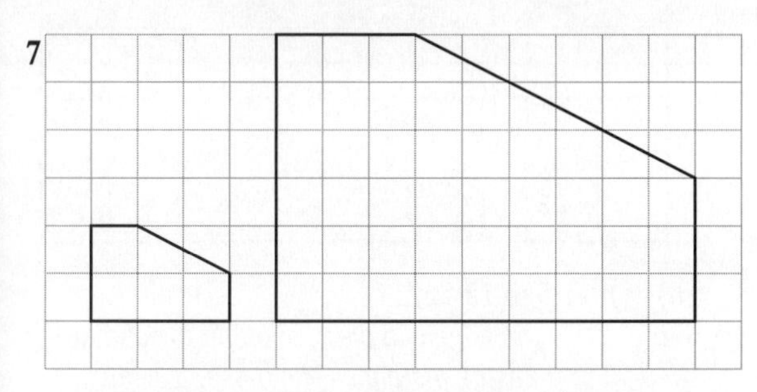

Since you are not told the centre of the enlargement, you can place your shape anywhere you want on the grid.

8 *Measure the distance of each point from the centre, P. Because the scale factor is 2, double each distance, and **measure from the centre again**.*

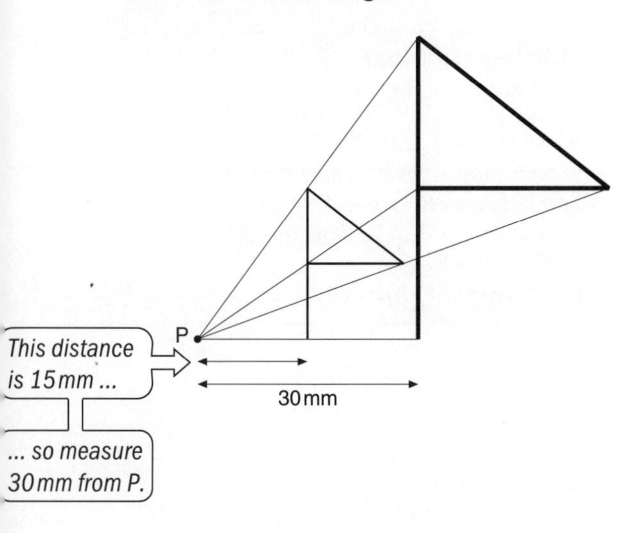

9 *There are lots of ways you could do this. Here are two examples.*
*The important thing is that there must be a **repeating pattern** to what you do and there must be no gaps. Just fitting the shapes on the grid randomly will not gain you the marks.*

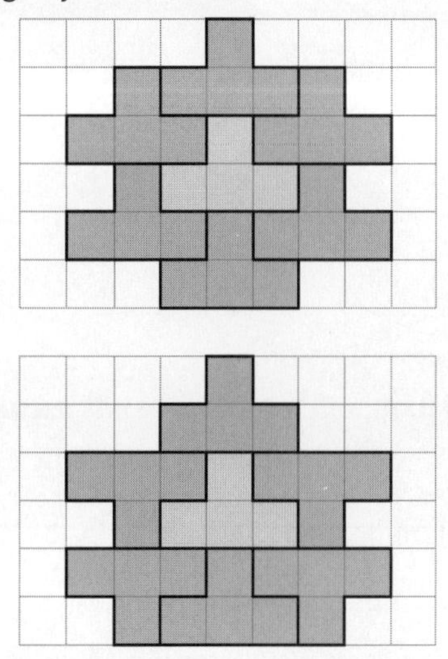

More help or practice

Drawing reflections and rotations
► Book B+ pages 24 to 26, 39 to 42

Enlargements and scale factors ► Book G3 page 7;
Book B4 pages 39 to 42, 52 to 53

Congruent figures ► Book G8 pages 80 to 81, Book G+ page 9

Scales and bearings (page 54)

1 (a) Tewkesbury

(b) Evesham

(c) Ledbury

Remember the grid reference is always to the bottom left-hand corner of the square.

(d) 2776

2 *If it helps, don't be afraid of turning the map round to check on left and right.*

(a) Penn Close

(b) Lyford Way

(c) Come out of Hunter Close, turn right, then left and left again. Go on to the roundabout and turn right.

3 (a) *Drawn half-size*

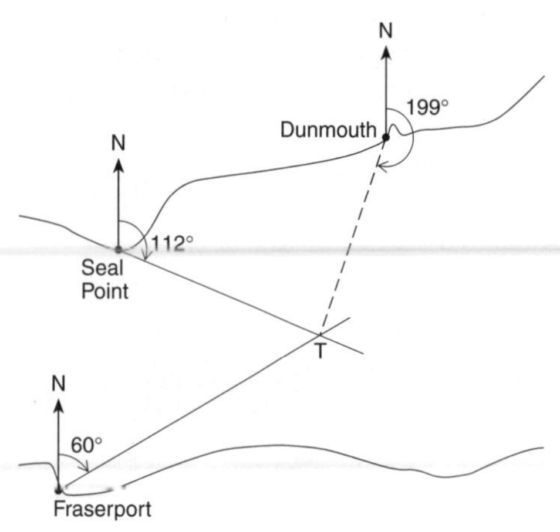

(b) 10·0 km *You should be within 0·5 km.*

(c) 199° *You should be within 2°.*

4 (a) and (b) *Drawn half-size*

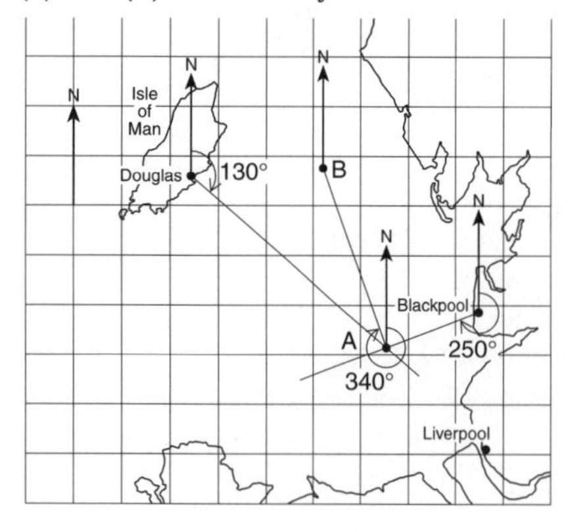

(c) (i) 3·0 cm (ii) 45 km (iii) 260° to 270°

If your answers are very different to these, check that you have got B in the correct position.

5

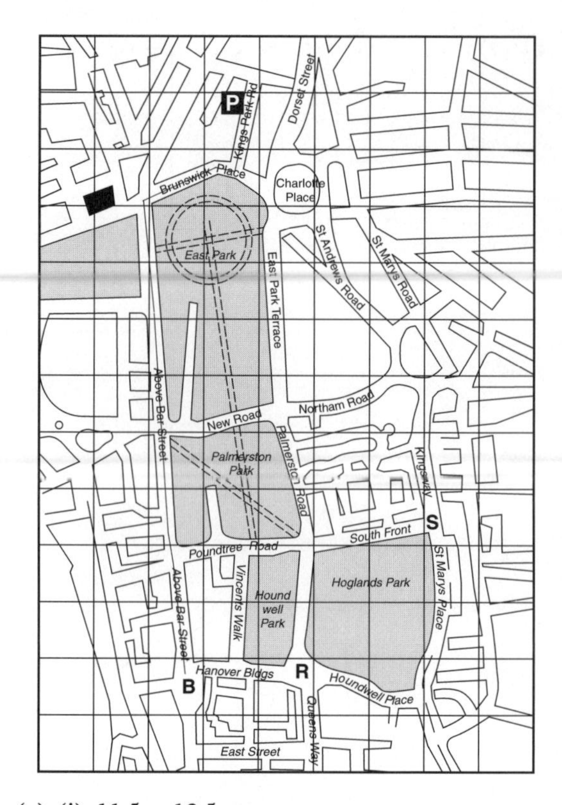

(a) (i) 11·5 to 12·5 cm

Use the edge of a piece of paper to estimate or measure distances like these. It is not sensible to give too 'accurate' an answer.

(ii) 1150 to 1250 metres

(iii) 1·15 to 1·25 km

(b) (i) 4·5 to 5·5 square cm

(ii) 45 000 to 55 000 m^2

(c) North-east

More help or practice

Interpreting maps, plans and photos
► Book G5 pages 18 to 25, Book G6 pages 44 to 49, Book G8 pages 10 to 13, Book B5 pages 36 to 39

Grid references ► Book G6 pages 15 to 17, Book B2 pages 6 to 11, Book B5 page 44

Scale drawings ► Book G8 pages 58 to 59, Book B3 pages 56 to 62

Bearings ► Book G8 pages 60 to 61, Book B3 pages 45 to 48

Units and measures (page 56)

If you are asked for an approximate answer don't give too many figures.

Where you are asked to give a rough answer, we have given limits, such as 1·4 to 1·6 metres. Your own answer should be between the limits.

1 (a) 3 litres is about 3×2 pints
$= 6$ pints. (5 to 6 pints)

(b) 2lb is about $2 \div 2$kg
$= 1$kg. (0·9 to 1kg)

2 20 feet is about $20 \div 3$ metres. (6 to 7m)

3 2200 feet is about $2200 \div 3$ metres
$= 733 \cdot \ldots$ metres. (650 to 740m)
733m is too accurate an answer.
It would be better to give 700m or 730m as your answer.

4 (a) 40lb is about $40 \div 2$kg
$= 20$kg. (18 to 20kg)

(b) 2 inches is about 2×25mm
$= 50$mm. (45 to 55mm)

(c) 8 gallons is about $8 \times 4{\cdot}5$ litres
$= 36$ litres. (30 to 40 litres)

(d) 5 miles is about 8 kilometres.
400 miles is 80×5 miles, so 400 miles is about 80×8km $= 640$km.
(600 to 700km)

5 (a) 5 litres *(5 litres is about 9 pints)*

(b) 100 inches *(100 inches is about 2·5 metres)*

(c) 1 foot *(1 foot is about 300 millimetres)*

(d) 1 pound *(1 pound is about 500 grams)*

6 (a) 15000 metres $= 15000 \div 1000$km
$= 15$ kilometres

(b) 0·5 metres $= 0{\cdot}5 \times 100$ centimetres
$= 50$ centimetres

(c) 150ml $= 150 \div 1000$ litres
$= 0{\cdot}15$ litres

(d) 1·5kg $= 1{\cdot}5 \times 1000$ grams
$= 1500$ grams

7 1·6km $= 1{\cdot}6 \times 1000$m $= 1600$m
So 1·6km + 600m = 1600m + 600m
= 2200 metres (or 2·2km)

8 1 litre is 1000ml, so you can serve
$1000 \div 25$ measures $= 40$ measures.

9 4kg is 4000g, so there are
$4000 \div 5 = 800$ sheets.

10 1·5 litres $= 1{\cdot}5 \times 1000$ml $= 1500$ml
So there is 1500ml + 300ml = 1800ml
= 1·8 litres in the jug.

11 (a) 30m $\div 2{\cdot}5$m/s $= 12$ seconds

(b) 18 sec $\times 2{\cdot}5$m/s $= 45$ metres

12 The car goes 5km in 2 minutes, so in 1 hour (60 minutes) it will go 30 times as far
= 150km.
So its speed is 150km/h.

13 16 minutes is 16×60 seconds $= 960$ seconds, so the speed is 1500m $\div$ 960 seconds
$= 1{\cdot}56\ldots$ m/s $= 1{\cdot}5$ or $1{\cdot}6$m/s.

14 (a) 150 beats $\div$ 3 minutes
= 50 beats per minute

(b) In 15 seconds it beats 20 times, so in 60 seconds (1 minute) it will beat 4 times as many times
= 80 beats per minute.

(c) In 45 seconds it beats 90 times, so in 15 seconds it will beat 30 times.
In 1 minute it will beat 4 times this
= 120 beats per minute.

15 (a) 1 minute is 60 seconds, so his speed is 400m $\div$ 60 sec $= 6{\cdot}66\ldots$ m/s
$= 6{\cdot}6$ to $6{\cdot}7$m/s

(b) In 20 seconds his heart beats 45 times, so in 60 seconds it will beat 3 times as many times
= 135 beats per minute.

16 (a) A points to 89·3.
B points to 95·6.

(b)

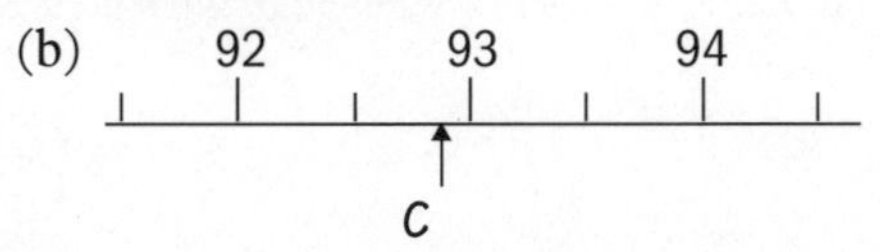

17 (a) (i) A points to 220.
B points to 180.
C points to 60.
On the scale there are 5 divisions to every 100.
So each division is 100 ÷ 5 = 20.

(ii)

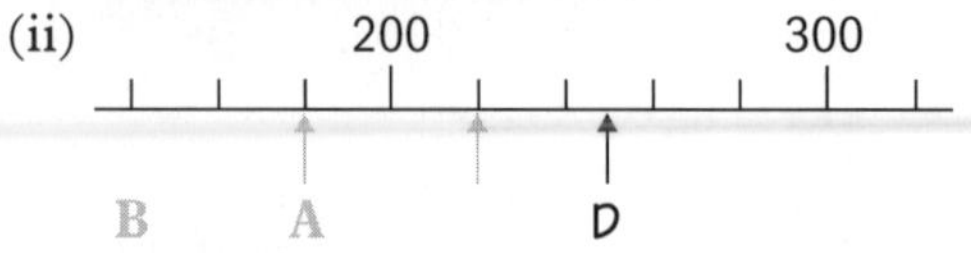

(b) (i) E points to 550.
F points to 375.
G points to 250.
On the scale there are 4 divisions to every 100.
So each division is 100 ÷ 4 = 25.

(ii)

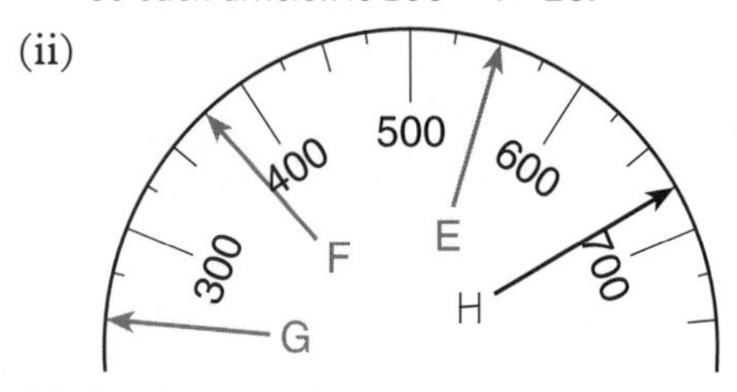

(c) I points to 1·4.
J points to 2·8.
K points to 0·8.
On the scale there are 5 divisions to every 1.
So each division = 1 ÷ 5 = 0·2.

18 8m
Any answer between 7m and 9m would be acceptable.
Mark the length of the bicycle on the edge of a piece of paper. Then see how many bike lengths fit into the bus length.

19 100 to 200 centimetres
A family car is about 135cm high.

20 (a) Height of mechanic is 1·5 to 2·0m.
A fairly tall man is about 1·8m.
Length of bus is 7·5 to 12m.

(b) *You could explain how you used a piece of paper, as in question 18. Or you could say:*
The picture of the mechanic is 20mm;
the bus is 110mm, so the bus is
$110 \div 20 = 5{\cdot}5$ times the mechanic.
So the bus is $5{\cdot}5 \times 1{\cdot}8\text{m} = 9{\cdot}9\text{m}$.
Use your estimate of the height of the mechanic here.

More help or practice

Estimating length ► Book G1 pages 1 to 5, Book G+ page 24
Metric units of length, weight and capacity
► Book G3 pages 1 to 6, 36 to 42; Book G4 pages 7 to 10; Book G7 pages 50 to 55; Book G8 pages 5 to 6; Book G9 pages 28 to 32; Book B3 pages 10 to 11; Book B5 pages 92 to 95
Converting between 'old' and metric units
► Book G9 page 77, Book RB+ pages 22 to 23
Speed ► Book G7 pages 37 to 39, 43; Book B2 pages 69 to 70; Book B5 pages 1 to 5
Working out rates ► Book G8 pages 75 to 79, Book B5 pages 47 to 48
Reading scales ► Book G1 pages 6 to 11; Book G2 pages 8 to 11, 27 to 33

Length, area and volume (page 59)

1 (a) Area = $2\text{cm} \times 6\text{cm} = 12\text{cm}^2$
Perimeter = 2cm + 6cm + 2cm + 6cm = 16cm

(b) Another rectangle, such as
1cm by 12cm (perimeter 26cm), or
3cm by 4cm (perimeter 14cm)

2 (a) Area = $14{\cdot}4\text{cm} \times 2{\cdot}5\text{cm} = 36\text{cm}^2$

(b) (i) The square would be 6cm by 6cm.
(ii) Your drawing of a square with each side 6cm

3 (a) 20cm^2

(b) A rectangle with area 20cm^2, such as 1cm by 20cm, or 2cm by 10cm, or 4cm by 5cm

4 (a) (i) A square with sides 7m
A square 8m by 8m would need 64 tiles.
(ii) 49 tiles

(b) The rectangle must be 12m by 5m or 10m by 6m, if they do not cut any tiles.
Other rectangles such as 2m by 30m would not fit in the hall.

5 The top layer of cubes is 8 by 5 = 40, and there are 4 layers.
So there are 40×4 cubes altogether = 160cm^3.

6 Volume $= 35\,\text{cm} \times 22\,\text{cm} \times 5\,\text{cm}$
$= 3850\,\text{cm}^3$

7 (a) Volume $= 8\,\text{cm} \times 5\,\text{cm} \times 10\,\text{cm}$
$= 400\,\text{cm}^3$

(b) You can fit 2 mini-packs upwards, 3 across and 2 deep. So you can fit in $2 \times 3 \times 2 = 12$ mini-packs.
Be very careful with packing questions – make sure the small objects will really fit into the bigger one.

8 (a) Volume of pack $= 8\,\text{cm} \times 14\,\text{cm} \times 5\,\text{cm}$
$= 560\,\text{cm}^3$

(b) (i) Area of top $= 8\,\text{cm} \times 5\,\text{cm} = 40\,\text{cm}^2$
(ii) Area of front $= 8\,\text{cm} \times 14\,\text{cm} = 112\,\text{cm}^2$
Area of side $= 5\,\text{cm} \times 14\,\text{cm} = 70\,\text{cm}^2$
Total area (2 of each of the above)
$= 2 \times (40 + 112 + 70)\,\text{cm}^2 = 444\,\text{cm}^2$

9 (a) $50\,\text{cm} \times 35\,\text{cm} = 1750\,\text{cm}^2$

(b) Area of side $= 20\,\text{cm} \times 35\,\text{cm} = 700\,\text{cm}^2$
Area of base $= 20\,\text{cm} \times 50\,\text{cm} = 1000\,\text{cm}^2$
Total area (2 sides, back and front and base)
$= (2 \times 700)\,\text{cm}^2 + (2 \times 1750)\,\text{cm}^2 + 1000\,\text{cm}^2$
$= 5900\,\text{cm}^2$

(c) Depth $= 35\,\text{cm} - 5\,\text{cm} = 30\,\text{cm}$,
so volume $= 50\,\text{cm} \times 20\,\text{cm} \times 30\,\text{cm}$
$= 30\,000\,\text{cm}^3$

(d) 1 litre is $1000\,\text{cm}^3$, so there are 30 litres of water in the tank.
She adds 1 ml for each litre, so she adds 30 ml.

More help or practice

Areas of rectangles and shapes made from rectangles
► Book G3 pages 16 to 26, Book G9 pages 64 to 65, Book B1 pages 1 to 7, Book B3 pages 72 to 75

Volume of a cuboid ► Book G8 pages 1 to 7, Book G9 pages 90 to 91, Book B3 pages 6 to 8

Surface area of a cuboid ► Book B3 page 9

Estimating areas of irregular shapes
► Book G6 pages 18 to 19, Book B1 pages 19 to 25

Triangles and circles (page 62)

1 A has area $6\,\text{cm}^2$.
B has area $7{\cdot}5\,\text{cm}^2$.
The circle has radius 2 cm, so the area is $\pi \times 2^2\,\text{cm}^2 = \pi \times 4\,\text{cm}^2 = 12{\cdot}5\ldots\,\text{cm}^2$
So C has area of about $13\,\text{cm}^2$.

2 Your triangle with area $12\,\text{cm}^2$
(It might have base 6 cm and height 4 cm, for example.)

3 (a) Circumference $= \pi \times 420$ feet
$= 1319{\cdot}\ldots$ feet $= 1300$ to 1320 feet

(b) 1300 feet

4 (a) Perimeter (starting at the top right)
$= (30 + 12 + 30 + 12 + 30 + 12 + 30)\,\text{cm}$
$= 156\,\text{cm}$

(b) Area of triangle $= \frac{1}{2} \times \text{base} \times \text{height}$
$= \frac{1}{2} \times (12 + 12 + 12) \times 24\,\text{cm}^2$
$= 432\,\text{cm}^2$

(c) Area of arrow = area of triangle + area of rectangle below it
$= 432 + (30 \times 12)\,\text{cm}^2$
$= 792\,\text{cm}^2$

5 (a) Area $= \frac{1}{2} \times 30\,\text{m} \times 12\,\text{m} = 180\,\text{m}^2$

(b) Split the shape into two, like this:

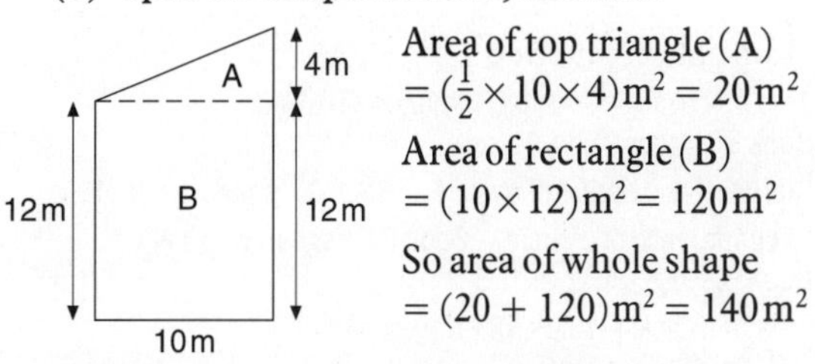

Area of top triangle (A)
$= (\frac{1}{2} \times 10 \times 4)\,\text{m}^2 = 20\,\text{m}^2$
Area of rectangle (B)
$= (10 \times 12)\,\text{m}^2 = 120\,\text{m}^2$
So area of whole shape
$= (20 + 120)\,\text{m}^2 = 140\,\text{m}^2$

6 Circumference of wheel $= \pi \times$ diameter
$= \pi \times 151\,\text{m} = 474{\cdot}38\ldots\,\text{m}$
There are 80 cars, so the space between them is $(474{\cdot}38\ldots \div 80)\,\text{m} = 5{\cdot}929\ldots\,\text{m}$
$= 6\,\text{m}$ (to nearest metre) or $5{\cdot}9\,\text{m}$ (to 1 d.p.)
It is not sensible to give an answer with more figures than these.

7 Suppose the circle has radius r m.
Area of circle = 20m^2,
so $\pi \times r^2 = 20$.
$r^2 = 20 \div \pi = 6{\cdot}366\ldots$
$r = \sqrt{6{\cdot}366\ldots} = 2{\cdot}523\ldots$
The largest circle he could paint has radius 2·5 m (to 1 d.p.).

8 Suppose the diameter is d metres.
Circumference of circle = 15 metres
so $\pi \times d = 15$.
$d = 15 \div \pi = 4{\cdot}777\ldots$
The largest diameter pen she could make is 4·8 m (to 1 d.p.).

9 (a) Radius of top = $10 \div 2$ inches
= 5 inches, so area of top
$= \pi \times 5^2 \text{ inch}^2 = 78{\cdot}5\ldots \text{ inch}^2$
$= 78 \text{ or } 79 \text{ inch}^2$.

(b) (i) Length of marzipan strip
= circumference of cake
$= \pi \times \text{diameter} = \pi \times 10$
$= 31{\cdot}415\ldots$ inches
Acceptable answers are 31 inches or 31·4 inches.

(ii) Area of strip = width × length
$= 3 \text{ inch} \times 31{\cdot}415\ldots \text{ inch}$
$= 94{\cdot}24\ldots \text{ inch}^2$
Acceptable answers are 94 inch² or 94·2 inch².

More help or practice
Area of a triangle ► Book G9 pages 66 to 70, Book B3 pages 76 to 81
Diameter and radius of a circle ► Book G7 pages 14 to 15
Circumference of a circle ► Book G7 pages 16 to 19, Book B2 pages 102 to 105
Area of a circle ► Book RB+ pages 73 to 78

Mixed shape, space and measures

(page 64)

1 (a) A is (0, 1), B is (⁻1, 3), C is (⁻2, 1) and D is (⁻1, ⁻1).

(b) Rhombus
ABCD is also a parallelogram, but 'rhombus' is the most informative name because ABCD has four equal sides.
Another word used for a rhombus is a diamond.

(c) Any six connected rhombuses similar to these will gain full marks.

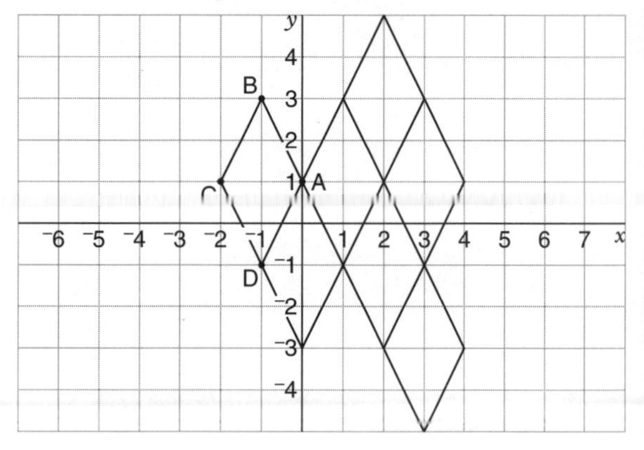

2 (a) Cube

(b) 2 cm

(c) One square has area 4cm^2, so
the total surface area $= 4\text{cm}^2 \times 6$
$= 24\text{cm}^2$

(d) Volume $= (2 \times 2 \times 2)\text{cm}^3$
$= 8\text{cm}^3$

3 (a) Number of discs $= 8 \times 6 = 48$
8 discs fit along one side and 6 discs along the other.

(b) Area of one disc $= \pi \times 2{\cdot}5 \times 2{\cdot}5\text{cm}^2$
$= 942\text{cm}^2$

(c) Area left $= [(30 \times 40) - 942]\text{cm}^2 = 258\text{cm}^2$

4 (a) The square measures 8 cm by 8 cm, so its area is 64cm^2.

(b) Area of triangle B $= \frac{1}{2} \times 8 \times 4\text{cm}^2$
$= 16\text{cm}^2$
Or you may have noticed it is a quarter of the square.

(c) Area of triangle D $= \frac{1}{2} \times 4 \times 2\text{cm}^2$
$= 4\text{cm}^2$

(d) Piece F is congruent to piece D.

5 (a) $x = 180° - 90° - 26° = 64°$
Angles on a straight line sum to 180°.
$y = 26°$
Z angles or alternate angles are equal.

(b) Area $= \frac{1}{2} \times 30 \times 12\text{cm}^2 = 180\text{cm}^2$

HANDLING DATA

Collecting data (page 66)

These are some possible answers to the questions.
Your answers may say the same thing but in a different way.
Your answers may be different but also be correct.

1 Ken's question
Advantage: students can state their favourite colour for the sweatshirt
Disadvantage: there may be many different answers and this could make the results for a large sample difficult to analyse.

Amy's question
Advantage: results for a large sample will be easy to analyse and display in a bar chart or table
Disadvantage: not a very wide range of colours – suppose you prefer yellow, for example

2 The question is too vague and would probably lead to a wide range of different responses that would be very hard to analyse.

3
- His survey would miss out people in full-time work.
- People using the bookshop may be less likely (or more likely) to use the public library.

4 *You could collect your results on one sheet such as the one below:*

TV watched on a Sunday

Name	Age	M/F	Minutes
1.			
2.			
3.			
4.			
5.			
6.			
7.			
8.			

OR
You could collect your results by using a questionnaire such as the one below.

TV watched on a Sunday

Name ______________________

1. Male ☐ Female ☐
2. How old are you?
   9–10 ☐ 11–12 ☐ 13–14 ☐ 15–16 ☐ 17–18 ☐
3. How many minutes TV did you watch last Sunday?
   0–59 ☐ 60–119 ☐ 120–179 ☐ 180–239 ☐ 240+ ☐

More help or practice
Surveys ► Book G9 pages 46 to 47, Book G+ pages 50 to 54, Book B+ pages 31 to 32

Timetables and calendars (page 67)

1 (a) 5:30 p.m.
If your answer is 7:30 p.m. it could mean that you counted the hours and minutes after 10:00 instead of 12:00 noon.
(b) 6:35 a.m.
(c) 9:15 p.m.
(d) 12:20 a.m.

2 (a) 25 minutes
(b) 15 minutes *19:45 is 7:45 p.m.*
(c) 75 minutes or 1 hour 15 minutes

3 (a) Friday
It can be helpful to count in 7's. For example, 1 July is a Wednesday so 8 July is a Wednesday too.
(b) Friday
There are 31 days in July.

4 *To follow a route, on some timetables you read across, on others you read downward. To follow the route for each train on this timetable you read downward.*

(a) 1 hour and 33 minutes or 93 minutes

(b) 14:32
If your answer is 14:54, you may have read across instead of downward.

(c) 12:20

> **More help or practice**
> a.m., p.m. and the 24-hour clock ► Book G3 pages 50 to 53
> Calculating with time ► Book G1 pages 33 to 38, Book G2 page 19, Book G5 pages 50 to 55, Book B4 pages 16 to 17, Book B5 pages 16 to 19
> Time and timetables ► Book G4 pages 20 to 27, Book B2 pages 24 to 28

Interpreting tables (page 68)

1 (a) Sue's hair is red.

(b) 4 people are female.
F stands for 'female'.

(c) David, Chris and Sue are under 21.

(d) Bill, David and Rashid are males with brown hair.

(e) $\frac{2}{8}$ or $\frac{1}{4}$ of the group wear glasses.
You do not have to write fractions in their simplest form unless you are asked to.

2 (a) It cost 26p to send a 50g letter first class.

(b) It cost 55p to send a 220g letter second class.

3 (a) It is 200 miles from Exeter to London.

(b) Tim travels further.
(Sarah travels 414 miles and Tim travels 456 miles.)

4 (a) 10 girls study 'A' level Mathematics.

(b) The number of students that study 'A' level English is $28 + 60 = 88$.

> **More help or practice**
> Reading tables ► Book G5 pages 32 to 35, Book B+ page 1
> Two-way tables ► Book B+ page 2
> Mileage charts ► Book G8 pages 14 to 15, Book B+ pages 3 to 4

Graphs and charts (page 69)

1 (a) Mehmet sold the most burgers.

(b) Tracy sold 18 burgers.

(c) They sold 116 burgers in total.
$(36 + 18 + 24 + 38 = 116)$

2

WEEKLY SALES	● represents 20 pizzas	Total
Monday	● ● ●	60
Tuesday	● ● ◔	45
Wednesday	● ● ● ●	80
Thursday	● ● ◖	50
Friday	○ ○ ○ ◔	75
Saturday	● ● ● ● ● ● ◖	130

3 (a) 30 red sweets

(b) 15 blue sweets

(c) $30 + 40 + 15 + 10 + 20 = 115$ sweets

4 (a) Most wells were drilled in 1990.

(b) 200 wells were drilled in 1988.

(c) 230 wells were drilled in 1991.
Each division on the vertical scale stands for 5 wells.
An answer of 206 means you may have counted in 1's from 200. An answer of 260 means you may have counted in 10's from 200.

5 (a) 7 girls have a shoe size larger than 3.

(b) The bars are shorter in the middle and taller at the left and right of the chart.
For a group of 12-year-old girls, it would be more usual for the bars to be taller around the middle of the chart and shorter at the left and right of the chart.

6

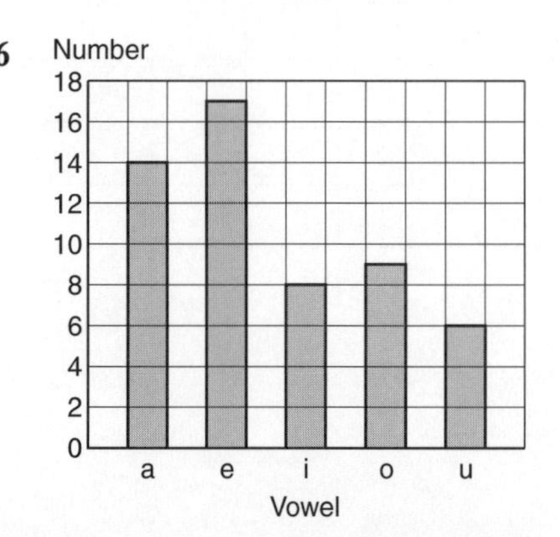

7 (a) Rainfall (mm)

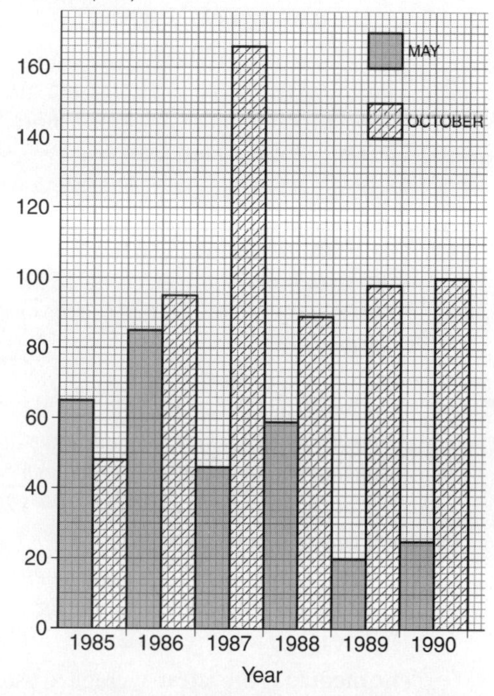

Each division on the vertical scale stands for 2 mm. When drawing the bars for 1988 and May 1990, be as accurate as possible.

(b) The May rainfall was greatest in 1986.

(c) The October rainfall was least in 1985.

(d) 1988

60 is exactly $\frac{2}{3}$ of 90 so
59 is approximately $\frac{2}{3}$ of 89.

8 (a)

Day of the week	Tally marks	Frequency
Monday (M)	卌 I	6
Tuesday (Tu)	III	3
Wednesday (W)	卌	5
Thursday (Th)	IIII	4
Friday (F)	卌 II	7
Saturday (Sa)	II	2
Sunday (Su)	III	3

You may find it helps to cross out the days lightly in pencil as you go along. If you make a mistake, you can rub out the marks and start again.

A useful check is to add up the frequencies and check the total is the same as the number of days in the list. The total is 30 which matches the number in the list.

(b)

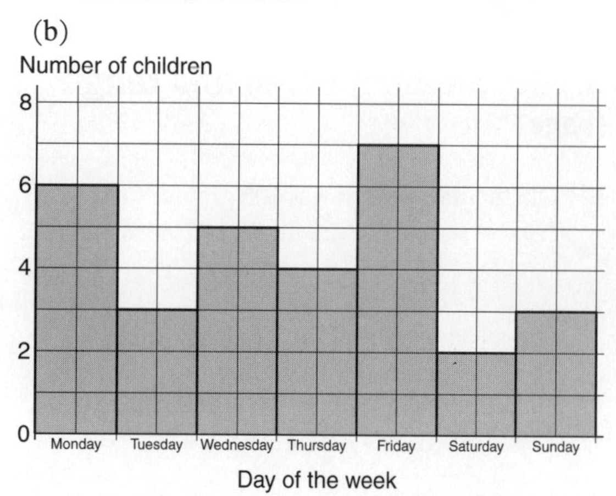

(c) Friday

9 (a)

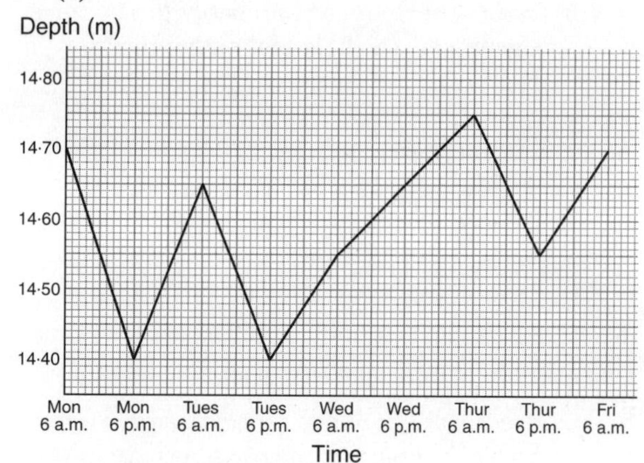

(b) About 14·55 m

(c) Monday

(d) Wednesday

From the shape of the graph for Monday and Tuesday it looks as though the level of water rises overnight and falls during the day as the water is used. The water level did not fall during the day on Wednesday which suggests heavy rain in the afternoon.

More help or practice

Bar charts and stick graphs ► Book G1 page 21, Book B5 pages 60 to 63

Line graphs ► Book G4 pages 12 to 13, Book B2 pages 118 to 119

Pictograms ► Book G+ page 47

Mode, mean, median and range
(page 72)

1 The median weight was 106 grams.
If your answer is 102 grams, it could mean that you did not put the weights in order first.

2 The mean weight of the Victoria plums is $(300 \div 10)$ grams = 30 grams.
The mean weight of the Merryweather plums is $(250 \div 10)$ grams = 25 grams.

3 (a) The mean weight of the members of the Hereward House team in kg is $616 \div 8 = 77$.

(b) Hereward House will probably win because their mean weight is a lot higher than Nelson House's mean weight.

4 (a) The mean weight in kg is $704 \div 8 = 88$.

(b) The range in kg is $99 - 80 = 19$.

5 The modal boot size is 8.

6 (a) The mean height in cm is $710 \div 10 = 71$.

(b) The median height is 71 cm.
If your answer is 69 cm, it could mean that you did not put the heights in order first.
If your answer is 70·5 cm, it could mean that you found the value half way between 70 and 71 instead of the value half way between 70 and 72.

(c) (i) 74 cm is the mode.
(ii) The mode is not a good measure of the average when there are few values.
As half the children are below 71 cm in height, 74 cm would not be a good measure of the average for these heights.

7 (a) (i) The range is $98 - 69 = 29$ years.
(ii) The mean age of the women is $1245 \div 15 = 83$ years.

(b) The oldest person in the home is a man.
The youngest man is 65 years or more.
The range for the men is 34 years.
So the oldest man is $65 + 34 = 99$ years or more.
The oldest woman is 98 so the oldest man is older than her.
Notice that you did not need to use the mean age of the men.

8 (a) (i) The median wage is £162·50
If your answer is £156, it could mean that you did not put the numbers in order first.
(ii) The mean wage is £2000 ÷ 10 = £200.

(b) (i) The median wage is £177.
If your answer is £150, it could mean that you did not put the numbers in order first.
(ii) The mean wage is £2002 ÷ 11 = £182.

(c) The median is the fairer average to use.
The mean wage for Bentgate is higher than the mean wage for Penlight because one person at Bentgate earns much more than the other people.
The medians show that half the people at Bentgate earn over £162·50 but about half the people at Penlight earn over £177.

9 (a) Leslie's mean score is $632 \div 8 = 79$.

(b) The range of Leslie's scores is $91 - 73 = 18$.

(c) Pat is a good choice as the mean score is the same but the range is less, showing less variation in his/her scores.
You could say that Leslie is a better choice because Leslie's highest score is better than Pat's. However, Leslie is a more risky choice.
Whatever choice you make, you have to back it up with good reasons.

10 (a) (i) The mean of the six remaining marks is $33 \cdot 6 \div 6 = 5 \cdot 6$.

(ii) The range of the six remaining marks is $5 \cdot 8 - 5 \cdot 3 = 0 \cdot 5$.

(b) It is probably fairer to count only the six remaining marks. When the highest and lowest marks are removed, the range is reduced to 0·5, showing that there is less variation in the marks now.

(c) If one of her marks is 5·9 and the range is 0·6, then her lowest mark must be 5·3 (5·9 – 0·6) or above. The mean cannot be less than the lowest mark.

More help or practice

Mean ► Book B3 pages 84 to 85, 87

Mean, median, mode ► Book G6 pages 50 to 52, Book G8 pages 62 to 65, Book RB+ pages 4 to 7

Range ► Book G8 pages 66 to 67, Book B3 page 86

Grouped data (page 75)

1 (a)

Number of hours watching television	Tally	Frequency
1–5	|	1
6–10	||||	4
11–15	~~||||~~ |	6
16–20	~~||||~~ ||||	9
21–25	~~||||~~ ||	7
26–30	||||	4

When making a tally chart, cross out the values lightly in pencil as you go along. If you make a mistake, you can rub out the marks and start again.

A useful check is to add up the frequencies and check it is the same as the number of values in the list. The total is 31 which matches the number of values.

(b) (i) 31 members

(ii) 20 members *(9 + 7 + 4)*

Grouped data and frequency diagrams (page 76)

1 (a)

Height (centimetres)	Tally marks	Number of pupils
140 and less than 150	|||	3
150 and less than 160	~~||||~~ |||	8
160 and less than 170	~~||||~~ ~~||||~~ ||	12
170 and less than 180	~~||||~~ |	6
180 and less than 190	|	1

Take care with 'borderline' values. For example, 160 is in the '160 and less than 170' group.

(b) 7 pupils were 170 cm or taller. *(6 + 1)*

2 (a)

Height (metres)	Tally	Frequency
$1 \cdot 40 \leq$ height $< 1 \cdot 45$		0
$1 \cdot 45 \leq$ height $< 1 \cdot 50$	||	2
$1 \cdot 50 \leq$ height $< 1 \cdot 55$	|	1
$1 \cdot 55 \leq$ height $< 1 \cdot 60$	|||	3
$1 \cdot 60 \leq$ height $< 1 \cdot 65$	~~||||~~ ||||	9
$1 \cdot 65 \leq$ height $< 1 \cdot 70$	~~||||~~ ||||	9
$1 \cdot 70 \leq$ height $< 1 \cdot 75$	||||	4
$1 \cdot 75 \leq$ height $< 1 \cdot 80$	||	2

Take care with 'borderline' values. For example, 1·60 is in the '$1 \cdot 60 \leq$ height $< 1 \cdot 65$' group.

(b) 6 women were shorter than 1·60 m.

3 (a)

Cost	Tally	Frequency
Up to £9·99	I	1
From £10 to £19·99	IIII	4
From £20 to £29·99	~~IIII~~ I	6
From £30 to £39·99	~~IIII~~ I	6
From £40 to £49·99	III	3

(b) 

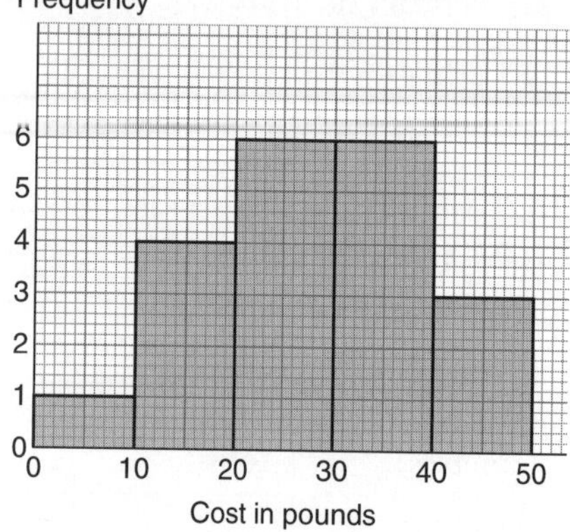

You need to decide on your own vertical scale. Find out how high the tallest bar will be and make sure your scale goes that far.

Try not to choose a scale that makes the bars too short. For example, the graph below is accurate but the bars are very short and the visual impact is lost.

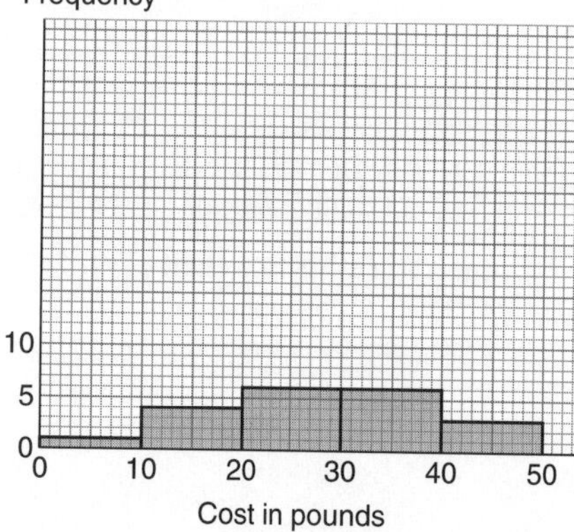

(c) 11 cameras cost less than £30. *(1 + 4 + 6)*

4 (a) (i) 6 grey squirrels weighed 600 g or more. *(4 + 2)*

(ii) 28 grey squirrels were weighed altogether. *(4 + 8 + 10 + 4 + 2)*

(iii) 'The modal group was: 500 g ≤ weight < 600 g'.

(b) (i)

Weight in grams	Tally marks	Frequency
200 ≤ weight < 300	~~IIII~~ III	8
300 ≤ weight < 400	~~IIII~~ ~~IIII~~	10
400 ≤ weight < 500	II	2

(ii)

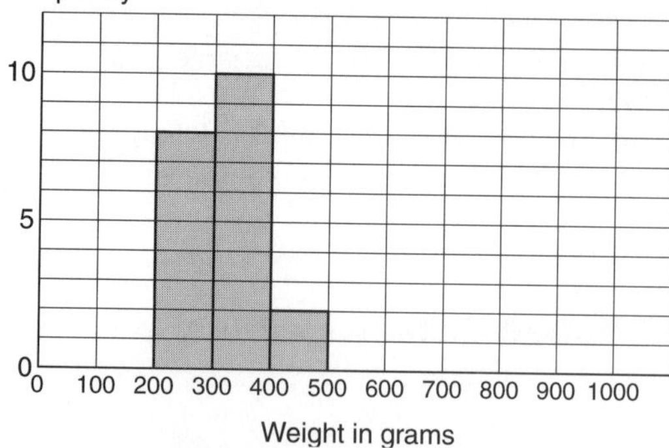

(c) Some possible comments are:

In general the grey squirrels are heavier – none of the red squirrels weigh more than 500 grams but 16 out of 28 grey squirrels weigh over 500 grams. The modal group for the grey squirrels was '500 g ≤ weight < 600 g' and the modal group for the red squirrels was '300 g ≤ weight < 400 g'.

There is less variation in the weight of the red squirrels – as the bar chart consists of five 'bars' for the grey squirrels but only three 'bars' for the red squirrels, the range for the red squirrels will be less.

5 (a) 4 girls qualified for a gold award.

(b)

Jump height (cm)	Tally	Frequency
12 and less than 16	||	2
16 and less than 20	|||	3
20 and less than 24	卌 |||	8
24 and less than 28	卌 |	6
28 and less than 32	||||	4
32 and less than 36	||	2

(c) 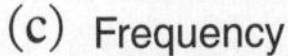

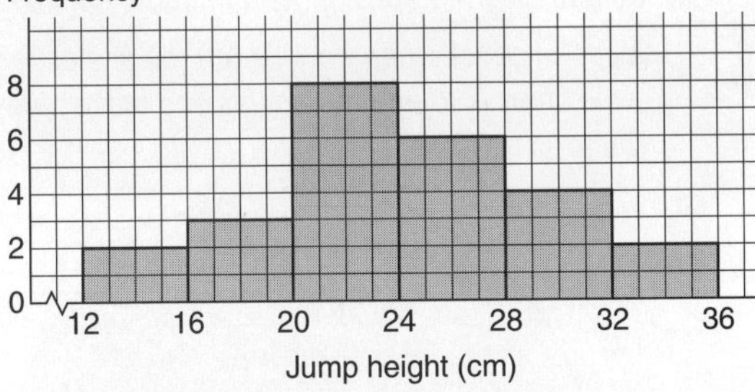

You may have chosen different scales.
You can show that a scale does not start at zero by breaking it as shown on the horizontal scale above.

(d) The frequency polygon is drawn on the frequency diagram.

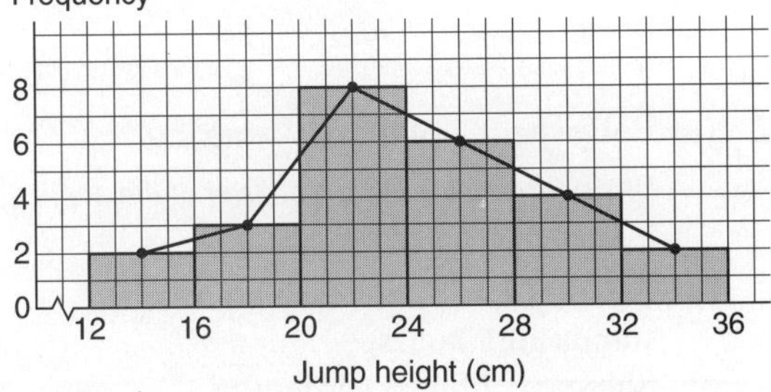

More help or practice

Grouped frequency diagrams ► Book G8 pages 28 to 33, Book B3 pages 82 to 84, Book RB+ pages 46 to 49

Pie charts (using a pie chart scale) (page 78)

You will lose marks if you leave the labels off the slices of a pie chart. Do not waste time in an examination shading your pie chart unless you are told to.

1 (a)

Source of income	Amount in millions	%
Membership	40·7	37
Inheritance	19·8	**18**
Investment	**18·7**	**17**
Rents	13·2	**12**
Other	17·6	16
Total	110·0	100

A useful check when working out percentages for a pie chart is to add them up and make sure the total is 100%.

(b)

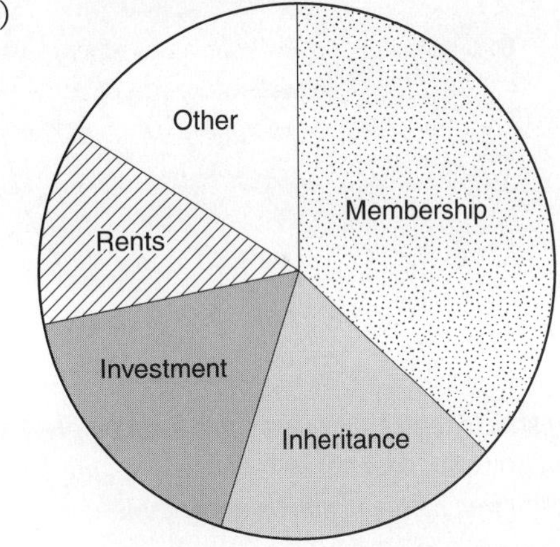

2 (a)

Place	Boys	%
Britain	110	44
France	75	30
Spain	15	6
Rest of Europe	40	16
Other	10	4
Total	250	100

An answer of 3% for France could mean you have interpreted 0·3 as 3%.

To convert decimals to percentages multiply by 100.

So 0·3 = 0·3 × 100% = 30%.

Remember to check for a total of 100%.

(b)

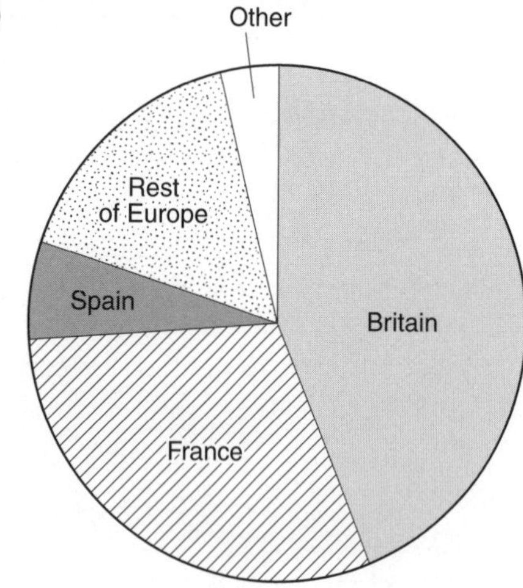

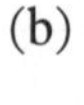

3 *You should read a pie chart scale as accurately as you can. However, your answers can vary slightly from those given here and still be acceptable.*

(a) 11% of the people who died were motor cyclists.

(b) The number of pedestrians killed in road accidents in 1993 was 33% of 3800 = 1254.

There are different ways to calculate 33% of 3800, for example:

0·33 × 3800 = 1254
(converting 33% to a decimal)

(3800 ÷ 100) × 33 = 1254
(finding 1% first)

(c) $\frac{5}{100}$ or $\frac{1}{20}$ of the people who died were cyclists.

Pie charts (using a protractor) (page 80)

1 (a)

Car colour	Number of cars	Pie chart angle
Red	55	165°
Blue	32	96°
White	20	60°
Grey	13	39°
Total	120	360°

360° ÷ 120 = 3° for each car. So each angle can be found by multiplying the number of cars by 3.

A useful check when working out the angles for a pie chart is to add the angles and make sure the total is 360°.

(b)

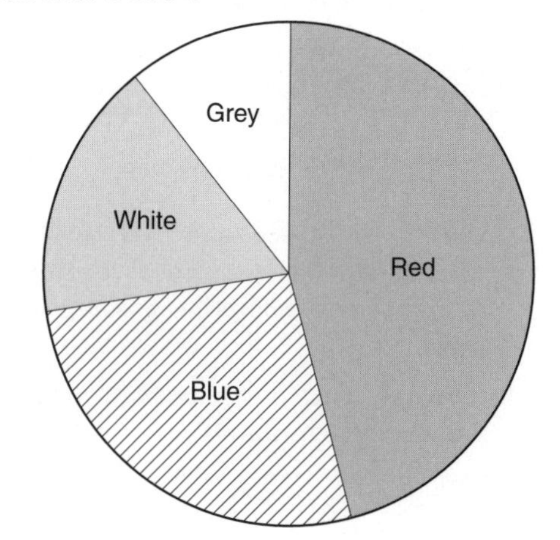

2 (a) Polly spent $\frac{1}{4}$ of the day working.

One right-angle is 90°, a quarter of the angle at the centre of the circle.

(b) 360° ÷ 24 = 15° for each hour, so 30° stands for 2 hours.
She spent 2 hours gardening.

(c) 60° stands for 4 hours so she spent 4 hours – 1 hour 30 minutes = 2 hours 30 minutes at her class.

(d) *Here are two ways of doing the calculation:*

- Work out the angle for 'sleeping' (360° – (60° + 60° + 30° + 90°) = 120°) and then calculate the number of hours.
  120 ÷ 15 = 8
- Work out the number of hours for each angle and take them away from 24.
  24 – (6 + 2 + 4 + 4) = 8

3 (a)

Method	Number of pupils	Pie chart angle
Bus	292	146°
Walk	180	90°
Car	120	60°
Cycle	62	31°
Train	66	33°
Total	720	360°

There are two ways to find the angles:

- *360° ÷ 720 = 0·5° gives the number of degrees for each pupil.*
  So now multiply the numbers of pupils by 0·5 to calculate the angles.
  (292 × 0·5 = 146°, 180 × 0·5 = 90°, ...)
- *720 pupils ÷ 360 = 2 pupils gives the number of pupils for each degree.*
  So now divide the numbers of pupils by 2 to calculate the angles.
  (292 ÷ 2 = 146°, 180 ÷ 2 = 90°, ...)

(b)

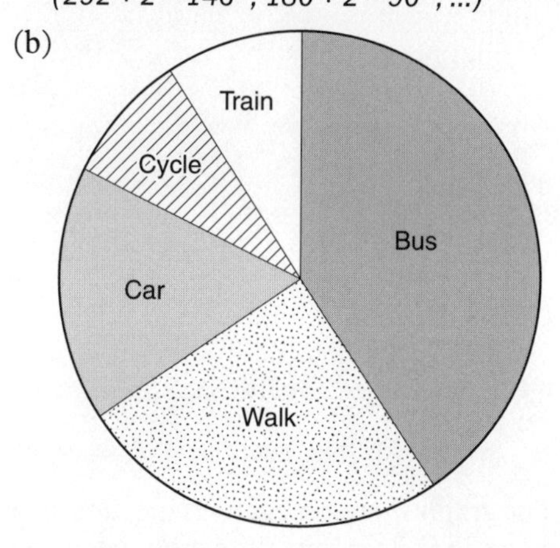

(c) In its simplest form, the fraction of pupils that come to school by car is $\frac{1}{6}$.

Write down all your working. You may get some credit for writing down a fraction equivalent to $\frac{1}{6}$.
Some fractions equivalent to $\frac{1}{6}$ are $\frac{120}{720}$, $\frac{60}{360}$, $\frac{6}{36}$, $\frac{3}{18}$.

(d) The percentage of pupils that walk to school is 25%.

4 (a) 800 women were at the concert.

Possibly the simplest way to calculate this is to work out how many people are represented by 1°:
1800 people ÷ 360 = 5 people
and then multiply to find the total number of women:
160 × 5 = 800 women

(b) (i) 600 ÷ 5 = 120 or 600 × 0·2 = 120
The angle for 600 men is 120°.

(ii)

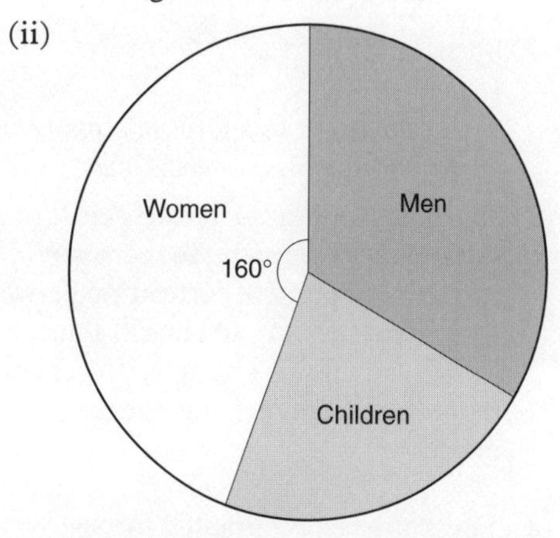

5 *You should read a protractor or angle measurer as accurately as you can. However, your answers can vary slightly from those given here and still be acceptable.*

(a) 36 × 0·25 = 9 or 36 ÷ 4 = 9
9 pupils were entered for Level 1.

(b) 20 ÷ 0·25 = 80° or 20 × 4 = 80°
The angle for Level 4 is 80°.

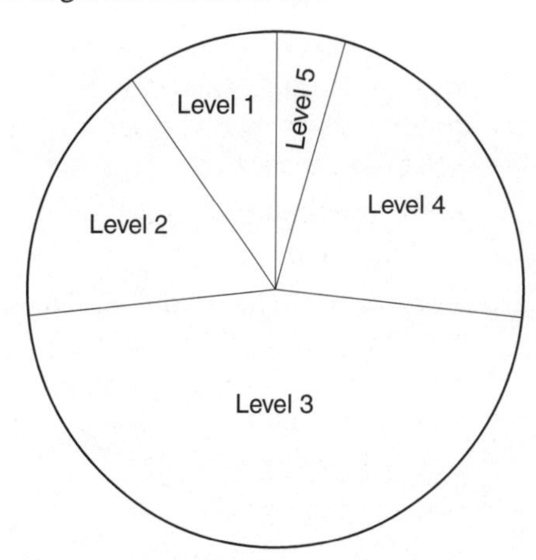

More help or practice

Reading pie charts ► Book G2 pages 54 to 56, Book G5 page 31
Reading and drawing pie charts ► Book G9 pages 71 to 76, Book B3 pages 52 to 53

Scatter diagrams (page 82)

1 The points appear to lie close to a line sloping down to the right, so there would appear to be negative correlation between temperature and rainfall. This means we can say that for example, 'In August, the higher the total rainfall, the lower the average temperature.'

2 (a) Paul: X
Spencer: Z
Liz: W
If you matched Liz with Y, you may have mixed up the time spent revising and the mark.

(b) The points appear to lie fairly close to a line sloping up to the right, so there would appear to be positive correlation between time spent revising and mark. This means we can say that for example, 'In general, the more time spent revising, the better your mark.'

3 (a) You would expect graph B for Sue.
You would not expect a connection between a person's height and their ability to score a good mark in a test.

(b) You would expect graph A for Aisha.
You would expect ice cream sales to be lower when there was more rain.

4 (a)

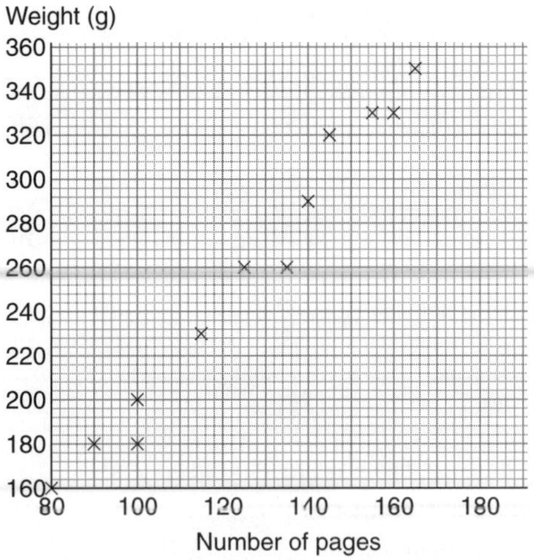

(b) The graph shows a positive correlation between the number of pages and weight. For these books, generally, the greater the number of pages, the heavier the book.

5 (a)

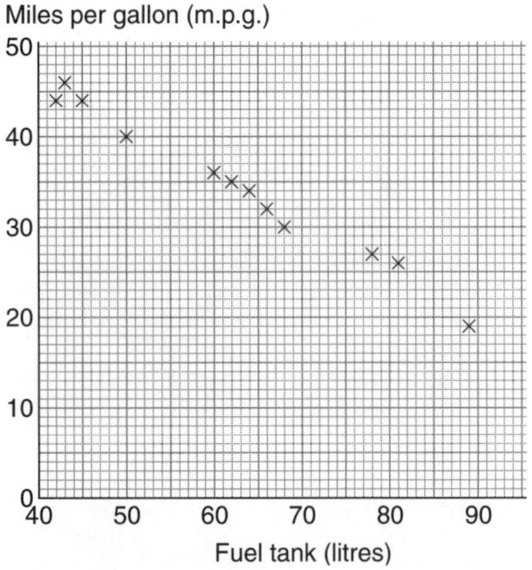

(b) The graph shows a negative correlation between fuel tank size and miles per gallon. For these cars, the larger the fuel tank the fewer miles per gallon.

(c) Any estimate from 27 to 29 m.p.g. is reasonable.

More help or practice
Scatter diagrams ► Book B3 page 89

Probability (page 84)

Probabilities should not be written as 'odds' or ratios, so do not write answers such as '1 in 4' or '1:3'.

Unless told otherwise, write your probabilities in the same way as any in the question. So, if probabilities are written as fractions give your answer as a fraction. If probabilities are written as decimals/percentages give your answer as a decimal/percentage.

1 (a) $\frac{4}{8}$ or $\frac{1}{2}$

You do not have to give fractions in their simplest form unless asked.

(b) $\frac{8}{8}$ or 1

(c) 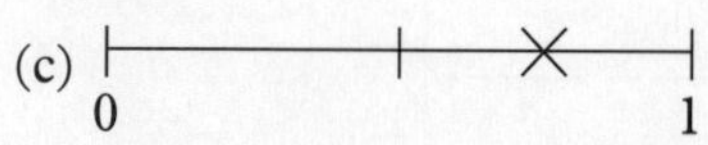

The probability of picking a number bigger than 3 is $\frac{6}{8}$ (or $\frac{3}{4}$). So the cross is placed $\frac{6}{8}$ of the way along between 0 and 1.

'Bigger than 3' does not include 3.

2 $\frac{2}{6}$ or $\frac{1}{3}$

'Less than 3' does not include 3.

3 $\frac{49}{50}$

4 (a) $\frac{1}{500}$ or 0·002

(b) $\frac{5}{500}$, $\frac{1}{100}$ or 0·01

The fractional form is more usual in this sort of question.

5 $1 - 0{\cdot}86 = 0{\cdot}14$

6 (a) There are $(4 + 10 + 9 + 5) = 28$ pupils in Form 12TY.

(b) (i) There are 14 girls in total so the probability that a pupil chosen at random is a girl is $\frac{14}{28}$ or $\frac{1}{2}$.

(ii) 15 pupils have fair hair so the probability that a pupil chosen at random has fair hair is $\frac{15}{28}$.

(iii) The probability that a pupil chosen at random is a boy with fair hair is $\frac{5}{28}$.

7 (a) (i)

	Frequency
Win	11
Lose	43
Total	54

(ii) $\frac{11}{54} = 0{\cdot}204$ (to 3 d.p.)

Experimental probabilities are often given as decimals correct to 3 decimal places.

(b) You would expect to win about $100 \times 0{\cdot}204 \approx$ 20 times.

More help or practice

Probability ► Book G8 pages 41 to 46;
Book B2 pages 95 to 101; Book B4 pages 20 to 27, 31;
Book B5 page 123

All the possible ways (page 86)

1

P	A	T
P	T	A
A	P	T
A	T	P
T	A	P
T	P	A

2 (a)

T-shirt	Shorts
White	Blue
White	White
Red	Blue
Red	White
Green	Blue
Green	White

(b) $\frac{1}{6}$

3 (a)

Main course	Pudding
S	I
S	T
S	A
S	C
F	I
F	T
F	A
F	C
V	I
V	T
V	A
V	C

(b) (i) $\frac{1}{12}$ (ii) $\frac{4}{12}$ or $\frac{1}{3}$ (iii) $\frac{9}{12}$ or $\frac{3}{4}$

4 (a)

Dice

Spinner		1	2	3	4	5	6
	1	1	2	3	4	5	6
	2	2	4	6	8	10	12
	3	3	6	9	12	15	18

(b) (i) $\frac{2}{18}$ or $\frac{1}{9}$ (ii) $\frac{8}{18}$ or $\frac{4}{9}$ (iii) $\frac{7}{18}$

5 (a) The possible outcomes are:

1, 1 1, 2 1, 3 1, 4
2, 1 2, 2 2, 3 2, 4

where the first number is from the grey set and the second from the white set.

You could show the possible outcomes in a table like this:

White card

Grey card		1	2	3	4
	1	1, 1	1, 2	1, 3	1, 4
	2	2, 1	2, 2	2, 3	2, 4

(b) $\frac{2}{8}$ or $\frac{1}{4}$

(c) $\frac{2}{8}$ or $\frac{1}{4}$

6 (a) Possible ways to score a total of 7 are:

1, 6 2, 5 3, 4 4, 3 5, 2 6, 1

where the first number is on the red dice and the second on the blue dice.

You could show the possible ways in a table like this.

Red dice	Blue dice
1	6
2	5
3	4
4	3
5	2
6	1

(b) There is only one way to score 2: 1 on the red dice and 1 on the blue dice. There are six ways to score 7, so a score of 7 is more likely.

More help or practice

Counting possible ways ► Book G3 pages 29 to 33, Book B5 pages 75 to 83

Counting possible ways and probability ► Book B4 pages 27 to 30

Misleading graphs (page 88)

1 (a) Diagram B is misleading.

(b) The sizes of the different sections are not correct. For example, the section representing protein (20%) should be twice as large as the section representing treats (10%) but it is much more than twice as large.

2 *Castlefield*

The graph suggests that from about 1985 the number of pupils increased a great deal. However, the number of pupils in 1985 was about 655 and in 1995 about 670. This shows an increase of only about 15 pupils in 10 years. The left-hand scale starts at 640 and this leads to the misleading impression.

Heathton

The height of the bars suggests that the number of pupils in1995 was about 6 times the number of pupils in 1965. However, the figures show that the number of pupils rose from 990 to 1200, and 1200 is far less than 6×990!

More help or practice

Misleading graphs and charts ► Book B5 pages 64 to 66

Mixed handling data (page 89)

1 (a) (i) Size 9 (ii) 62 pairs

(b) Sizes 3 and 12

(c) A jagged line has been used to show that the scale does not start at 0.

(d) A dotted line has been used to show that not all points on the line have meaning. For example, there are no trainers sold that are size $3\frac{1}{4}$.

2 (a) 56 minutes (b) 1422 (c) 1629

3 (a) (i) Table C

(ii) There are six faces and each has an equal change of appearing. So in 120 rolls, each face would appear about $120 \div 6 = 20$ times.

Since there are 2 red faces, you would expect red about $20 \times 2 = 40$ times. For the same reason you would expect blue about 40 times.

You would expect yellow and green about 20 times.

Table C is closest to these figures.

(b) 2000

4 (a)

	Frequency
Win	18
Lose	82

(b) $\frac{18}{100}$ or 0·18

*The probability is not $\frac{18}{82}$ as 82 is not the **total** number of rolls.*

(c) $0{\cdot}18 \times 250 = 45$

5 (a) (i) $100 + 900 + 200 + 300 + 100 = 1600$
(ii) Castle Walk
(iii) Castle Walk
(b) (i) Parson's Walk
(ii) $20 \leq$ age < 40
(c) (i) Washington Heights
(ii) Parson's Walk
(iii) Castle Walk
(d) Any fraction equivalent to $\frac{600}{1200}$, for example: $\frac{6}{12}$, $\frac{1}{2}$.
(e) (i) $\frac{900}{1600}$ $(= \frac{9}{16})$ or 0·5625 or 56·25%
(ii) $\frac{400}{1600}$ $(= \frac{1}{4})$ or 0·25 or 25%

6 (a) Alan's mean score is $153 \div 9 = 17$.
(b) The range of Alan's scores is $26 - 8 = 18$.
(c) (i) Alan has played more games than Dharmesh so it is not fair.
(ii) He should choose Dharmesh. Their ranges are the same, but Dharmesh's mean score is higher than Alan's.

7 (a) (i) TRUE (ii) FALSE (iii) TRUE
(b) You could choose some more schools that are not near Ashurst.
For each school, carry out a survey on enough cars, say 100.
Draw a bar chart of the percentages using the same scale as the Ashurst one.
Compare the bar charts and decide if they have the same shape.

More help or practice
Mixed handling data ▸ Book B5 pages 122 to 123

MIXED QUESTIONS

Mixed questions 1 (page 93)

1 (a)
(b)

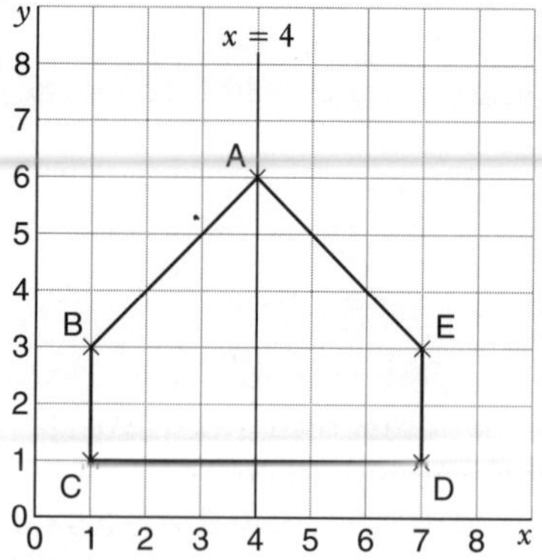

(c) (7, 3)
(d) $x = 4$

2 (a) £17·05
(b) £15·30
Start by finding the weight of the parcel (14·5 kg).

3 1 m 8 cm, 1·12 m, 150 cm, 1·63 m, 1·70 m

4 (a) 3:00 p.m.
(b) 70 miles
(c) 5 hours
(d) $\frac{200 \text{ miles}}{5 \text{ hours}} = 40$ m.p.h.

5 (a) $(2 \times 4) - 1 = 7$
(b) An input of 4 gives an output of 15, which is *not* a prime number. An input of 7 gives an output of 27, which is also not a prime number.
There are other inputs which show that Paul is wrong, for example 9 and 10.
(c) $4n - 1$

6 (a) (i) 36
(ii) 8, 16, 24, 32, 40
(iii) 2, 5, 7
(iv) $3^3 = 27$ and $\sqrt{25} = 5$
(b) (i) 23, 27
(ii) Add 4 to the number before.
(iii) $4n - 1$

7 (a) Volume $= (6 \times 6 \times 12)\text{cm}^3$
$= 432\text{cm}^3$

(b) $1296 \div 432 = 3$
The Economy Size holds 3 times more tea than the Standard Size packet.
The weight of tea in the Standard Size packet is 110g.
So the weight of tea in the Economy Size packet is $110\text{g} \times 3 = 330\text{g}$.

8 (a) *Remember that 1 km = 1000 m.*
The number of lengths $= 1000 \div 25 = 40$

(b) (i) 29 is roughly equal to 30 and £1·95 is roughly equal to £2.
So Cala will get roughly £2 × 30 = £60.

(ii) £60 must be bigger than the exact amount because both numbers are rounded up.

Mixed questions 2 (page 96)

1 (a) 800g (b) 3 (c) 6:10 p.m.

2 (a) $V = d \times 5 \times 4$
$= 20d$

(b) (i) $5d$ (ii) $4d$

(iii) The surface area of the cuboid is equal to the sum of the areas of the 6 rectangles.
$20 + 4d + 5d + 4d + 5d + 20 = 40 + 18d$

3 (a) $2\frac{1}{2} + 1\frac{3}{4} + 1\frac{1}{2} = 2\frac{1}{2} + 1\frac{1}{2} + 1\frac{3}{4}$

$= 4 + 1\frac{3}{4} = 5\frac{3}{4}$

Sara sold $5\frac{3}{4}$ yards.

(b) $10 - 5\frac{3}{4} = 10 - 5 - \frac{3}{4}$

$= 5 - \frac{3}{4}$

$= 4\frac{1}{4}$

$4\frac{1}{4}$ yards were left.

(c) $3\frac{1}{2}$ yards $= (3 \times 36) + (\frac{1}{2}$ of 36) inches
$= (3 \times 36) + (36 \div 2)$ inches
$= 108 + 18$ inches
$= 126$ inches

Albert had 6 inches to spare. *(126 − 120 = 6)*

4 (a) 111 minutes

(b) Susan travels 30 miles in 60 minutes, so she travels 40 miles in 80 minutes.
She arrived in Cardiff at 13:20.

5 (a) *To find the area of the ground floor, divide the plan into two rectangles. You can do this in two ways.*

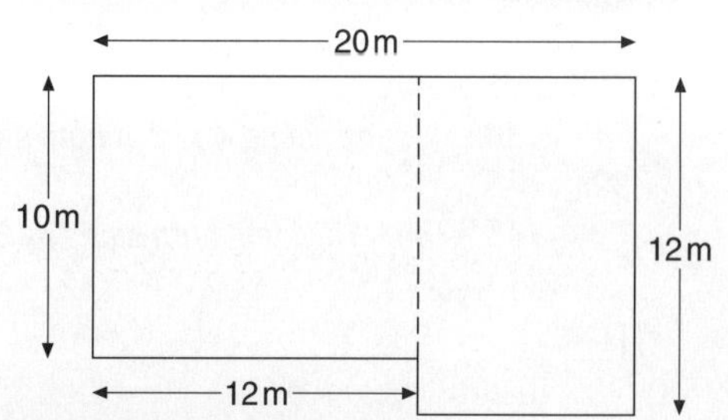

Total area in square metres
$= (10 \times 12) + (12 \times 8)$
$= 120 + 96$
$= 216$

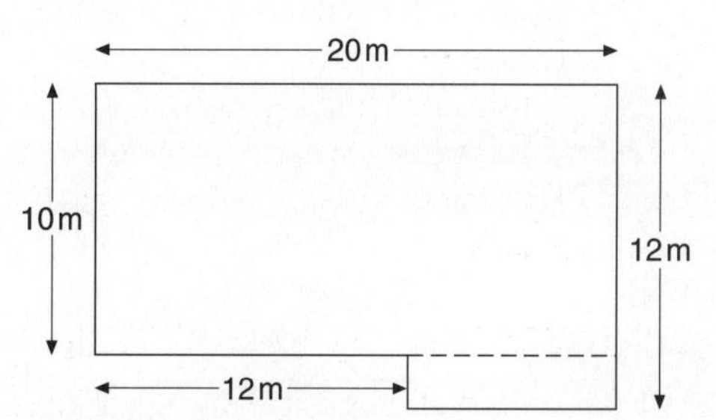

Total area in square metres
$= (10 \times 20) + (2 \times 8)$
$= 200 + 16$
$= 216$

(b) Cost of building = £225 × 216
= £48600

(c) 6% of £48600 = £48600 × 0·06
= £2916

Mixed questions 3 (page 98)

1 (a) Cost in pounds $= 45 + (30 \times \frac{1}{2})$
$= 45 + 15$
$= 60$

(b) *One way to work this out is to form and solve an equation.*
Let n be the number of hours worked. Then
$90 = 45 + 30n$
$45 = 30n$ *(subtracting 45 from both sides)*
$30n = 45$
$n = 45 \div 30$ *(dividing both sides by 30)*
$= 1{\cdot}5$ or $1\frac{1}{2}$

2 (a) 30% of 300 $= 300 \times 0{\cdot}3$
$= 90$

(b) 11% is the same as 0·11.
So the angle of the pie chart
$= 360° \times 0{\cdot}11$
$= 39{\cdot}6°$
$= 40°$ (to the nearest degree)

(c) 22% = 0·22

(d) $12\% = \frac{12}{100} = \frac{3}{25}$

3 (a)

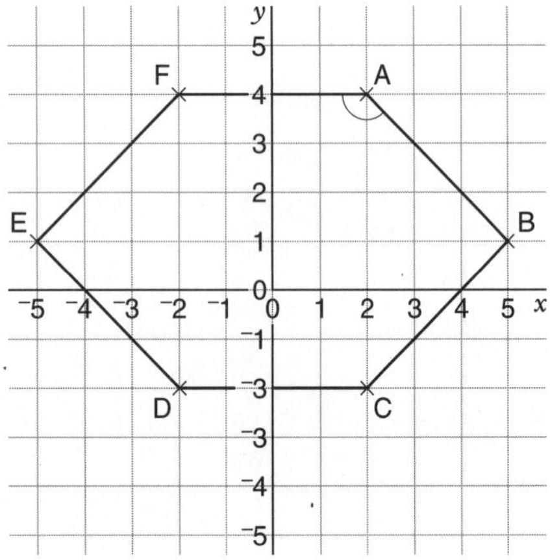

(b) Hexagon

(c) The angle at A is marked, but the angles at C, D and F are also obtuse.

(d) 2

4 (a) 1 kg = 1000 g
100 g of flour makes 15 biscuits, so
1000 g of flour makes 150 biscuits.

(b) 160 g of flour

5 (a) 136 km + 180 km = 316 km

(b) 8 km ≈ 5 miles
1 km ≈ $\frac{5}{8}$ mile
So 136 km ≈ $136 \times \frac{5}{8}$ miles
= 85 miles

(c) The distance to Dijon is 180 km.
Andrew travels 80 km in 1 hour,
160 km in 2 hours and
180 km in 2 hours 15 minutes.

6 (a) Volume $= (11 \times 3{\cdot}5 \times 6)\,\text{cm}^3$
$= 231\,\text{cm}^3$

(b) 1·6 kg = 1600 g
Percentage of marzipan
$= \frac{250}{1600} \times 100\%$
$= 15{\cdot}625\%$
$= 15{\cdot}6\%$ (to 1 d.p.)
$= 16\%$ (to the nearest whole number)
It is important to work with the same units. The working above uses grams but you could have changed 250 g to 0·25 kg and worked in kilograms. In this case the calculation would have been $\frac{0{\cdot}25}{1{\cdot}6} \times 100\% = 15{\cdot}6\%$ which gives the same answer.

(c) Area of top of cake $= 20 \times 25\,\text{cm}^3$
$= 500\,\text{cm}^2$
Thickness of marzipan $= 231 \div 500\,\text{cm}$
$= 0{\cdot}462\,\text{cm}$ or $0{\cdot}46\,\text{cm}$